THE UNIVERSE AND THE ATOM

THE UNIVERSE AND THE ATOM

Don Lichtenberg

Indiana University, Bloomington, USA

NEW JERSEY · LONDON · SINGAPORE · BEIJING · SHANGHAI · HONG KONG · TAIPEI · CHENNAI

Published by

World Scientific Publishing Co. Pte. Ltd.
5 Toh Tuck Link, Singapore 596224
USA office: 27 Warren Street, Suite 401-402, Hackensack, NJ 07601
UK office: 57 Shelton Street, Covent Garden, London WC2H 9HE

British Library Cataloguing-in-Publication Data
A catalogue record for this book is available from the British Library.

THE UNIVERSE AND THE ATOM

ISBN-13 978-981-270-606-5
ISBN-10 981-270-606-2
ISBN-13 978-981-270-561-7 (pbk)
ISBN-10 981-270-561-9 (pbk)

Printed in Singapore.

To Rita, Naomi and Rebecca

Preface

One thing that distinguishes human beings from the other animals is our curiosity about nature. A wolf may look at the moon and howl at it, but only human beings want to know how big it is and how far away, what it is made of and why it changes its appearance. More generally, we want to know the way nature works, even if our knowledge brings no practical benefit. As history has turned out so far, some of the knowledge we have gained about nature has not helped us materially at all, but other knowledge has led to profound changes in our civilization. Unfortunately, not all these changes have been for the good.

Our senses let us perceive nature only at intermediate sizes, but not at the very small and the very large. We can look at a tree but we cannot see the atoms that make up the tree. We can look at a portion of the earth, but our vantage point does not let us see directly that the earth is a sphere.

In this book we shall be principally concerned with nature at both the smallest and the largest scales, from atomic and subatomic particles to the universe as a whole. We want to answer as well we can the questions: "What are we made of?" and "What is the nature of the universe in which we live?"

We are, of course, made of flesh and blood, bones and hair, and we live on the earth. But on a smaller scale our bodies are made of cells, and the cells are made of molecules. In turn, the molecules are made of atoms, and atoms are made of electrons and atomic nuclei. As far as we know, the electrons are elementary in the sense that they are not composed of smaller things. However, atomic nuclei are made of protons and neutrons, which in turn are made of quarks. That is where our present understanding of the nature of matter ends, although there are speculations about even deeper levels. On a larger scale, the earth is the third planet from the sun in our solar system, and the solar system is in our galaxy, the Milky Way, which is a system of more than a hundred billion stars plus dust and other kinds of matter held subject to their mutual gravitational attraction. The Milky Way in turn is only one galaxy (system of stars) of many billions of galaxies that exist in the visible universe. And there have been speculations about what lies beyond the visible universe and whether we can ever know anything about it.

We are about to take a journey into the world of the very small and the world of the very large. What we shall find will be so foreign to our senses and so much against our intuition as to be almost beyond belief. Mankind has long thought about what matter is made of and what the universe is like. However, it was not until the twentieth century that our knowledge has grown so much that today we have a largely coherent picture at what nature is like at both large and small scales.

In order to make sense out of the very large and the very small, we have to devote some of our effort to examining nature on a human scale. In this way, we learn some of the laws of physics that we need, and we also learn of some laws that seem to satisfy common sense, but have only a limited range of validity. We often take a historical approach and discuss old ideas that seem either quaint or silly today, but it is beneficial to learn what went wrong with earlier scientific ideas.

Scientific knowledge is always provisional. Based on past experience, we can be confident that new discoveries will change the way we look at nature in many ways. However, many of our present ideas are rooted firmly in the results of a large number of observations and experiments, and so we can be confident that much of our present picture will endure.

The bibliography contains a number of books on the topics I treat here. I acknowledge freely consulting several of these books to obtain part of the material I used. I consulted other sources, especially on the web, as well. I have made decisions about what information to include, updated material when necessary, and put my own stamp on the presentation. I also acknowledge beneficial conversations with my good friends and colleages at Indiana University, Steven Gottlieb and Roger Newton. I am thankful to Bruce Carpenter of Indiana University for doing the drawings and to Andrew Chan Yeu Tong for his assistance with the editing and Alvin Chong of World Scientific for his cooperation in publishing this book.

Contents

Chapter 1

Early Ideas about the Universe

> To deride the hope of progress is the ultimate fatuity, the last word in poverty of spirit and meanness of mind.
>
> —Sir Peter Medewar (1915–1987)

1.1 The earth

In early times, most scholars regarded the earth as the center of the universe. This is only natural, as the earth appears huge to our senses. Everything in the heavens looks tiny by comparison, even the sun. In various cultures, the earth was regarded either as a flat disk or as a sphere. It is also understandable why some civilizations thought the earth was flat except for irregularities, such as mountains, on its surface. The reason, of course, is that the earth is so large that we cannot normally see the curvature of its surface. This is the first of many instances in which we shall find that the universe is quite different from the way it appears to our senses.

We can get a clue that the surface of the earth is curved. If we look at a ship approaching us far away at sea, the top of the ship comes into view first, when the part at the waterline is still below the horizon.

In modern times, we have had ample evidence that the earth is a sphere (to a good approximation). A few hundred years ago explorers circumnavigated the earth with sailing ships. More recently, airplanes and manned artificial satellites have flown around the earth. The fact that the earth looks round from every vantage point shows that it is a sphere. If the earth were some other shape, for example, a round disk, it could be viewed edgewise and appear to be thin. A sphere has more symmetry than any other shape because it is the only shape that looks round no matter from where it is viewed. Of course, the view has to be from far enough away that we can see the earth's roundness. The earth is so large that from its surface it appears flat, or rather, lumpy, with mountains and valleys.

It is interesting that many people who thought the earth was flat regarded it as a disk rather than a square or some irregular shape. Also, most of the flat-earth people thought they stood on the top side of the earth, the top being toward the heavens. What was below was only guessed at. People who believed in a flat earth thought that it was possible to fall off its edge, but, somehow, nobody ever saw the edge, and so they must have thought that they were always far away from it. The alternative view, that the flat earth was infinite in size (in other words, that there was no end to the earth in any direction) does not seem to have been seriously considered by most flat-earth advocates.

Those who regarded the earth as a sphere (apart from irregularities such as mountains) knew that it could be finite in size without having an edge. We now know that the earth is not quite a sphere, as it bulges outward at the equator because it is rotating on its axis once a day. With a spherical or almost spherical earth, the direction "down" is toward the center of the earth. People thought either that it was "natural" for bodies to fall toward the earth or that bodies fell because of some "influence" (later called gravity). There is not as much difference between these two points of view as appears, because gravity is a natural influence. Today we realize that a quantitative description of how bodies fall is very useful.

1.2 The heavens

According to many of the ancients, the heavens are what we see above the earth. These consist of the sun, the moon, the planets, the stars, and other objects occasionally, like comets and meteors. Even in early times, the planets were distinguished from the stars by their motion. They moved relative to the stars, and the name "planet" in Greek means "wanderer."

One of the earliest civilizations to have ideas about the universe was the Babylonian, located in what is now Iraq. We know from ancient writings that by 1700 BCE the Babylonians knew the length of a year to within a few minutes. The Babylonians divided the year into twelve lunar months. A lunar month is the time for the moon to go from full moon to full moon, as observed on earth, a time of somewhat more than 29 days. Because a year is a little longer than 12 lunar months, the Babylonians included a leap month every few years.

The Babylonians divided the circle into 360 degrees, each degree into 60 minutes, and each minute into 60 seconds. We still use these units of angular measurement, but we also use others. In Figure 1.1 we show a cirlce with angles of 30 degrees and 90 degrees marked off. A degree is abbreviated °.

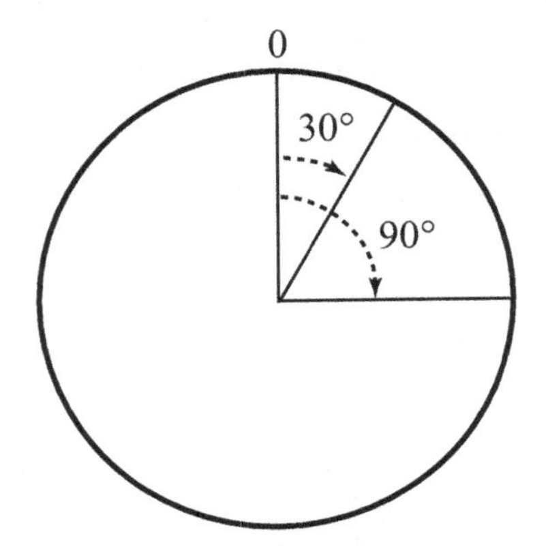

Figure 1.1: A circle, showing angles of 30 degrees and 90 degrees.

1.3 The views of Aristotle and Aristarchus

We skip over the accomplishments of the Chinese and Indian astronomers and pass directly to the Greeks. The most influential Greek philosopher was Aristotle (384–322 BCE). He proclaimed that knowledge of nature should be based on observation, an idea that was a great advance over that of Plato, who insisted that knowledge should come primarily from pure reason. However, Aristotle did not always follow his own precepts, often slipping into the patterns of thought of Plato. Furthermore, Aristotle often drew wrong conclusions from his observations.

Aristotle regarded the earth, the sun, and the moon as spherical, and said that the heavenly bodies move in circles around the earth, which is stationary and at the center of the universe. In this scheme, the sun and the stars go around the earth once every day, whereas we now know that the earth rotates around its axis once a day, giving rise to the apparent motion of the sun and stars.

Aristotle took over most of his ideas about the earth and the heavens from earlier Greek philosophers, including Pythagoras and Plato. Who can blame Aristotle? From our vantage point it certainly looks like the earth is the center of the universe.

The first Greek I know of who claimed the earth is spherical was Pythagoras (d. 497 BCE). Pythagoras did not give any evidence for his belief, but made his claim on esthetic grounds. On the other hand, Aristotle had several reasons for believing that the earth is spherical, the most compelling one being that during a lunar eclipse, the shadow of the earth appears as part of a circle as it moves across the surface of the moon.

Somewhat later, Eratosthenes (b. around 276 BCE, d. ?), who was born in what is now Libya and lived in Athens and Alexandria, made a measurement that enabled him to estimate the circumference of the earth. He had heard that on the summer solstice (June 21), the sun was directly overhead at noon in a place called Syene, now Aswan, Egypt. Eratosthenes, then living in Alexandria (which is north of Syene), found that the shadow of a stick made an angle of about 7 degrees, or about 1/50th of a complete circle. (The angle is the one made between the vertical and the line connecting the top of the stick to the end of the shadow.) He was able to find out the distance between between Syene and Alexandria, and calculated that the circumference of the earth was about 50 times as great. We are not sure of the exact size of the units of distance used by Eratosthenes, but he came pretty close to determining the circumference of the earth to be around 25,000 miles, its present value.

We return to Aristotle. In his view, the earth and its neighborhood are made of four elements: earth, air, water, and fire. The heavens are made of a fifth substance, which he called "ether" or "quintessence." Today we know that Aristotle was naive in his assertions. For example, there are many different substances in what he called "earth," as it is easy to tell by just looking at the large variety of different objects on the earth. We now know that these different objects are made of fewer than 100 naturally occuring substances called "elements." We shall discuss the elements in much more detail in later chapters.

The Greek astronomer Aristarchus of Samos (310–230 BCE), deduced the relative sizes of the earth, sun, and moon by geometrical means. Finding that the sun is much larger than the earth, he concluded that the earth must move around the sun, because he could not believe that the greater body moves around the lesser one. The notion that the earth revolves around the sun is called the "heliocentric" view.

Aristarchus also believed that the earth rotates on its axis, causing day and night, and that the earth's axis is tilted with respect to the ecliptic (the name given to the plane of the earth's orbit around the sun). We do not have the writings of Aristarchus to support his conclusions, but he is quoted by the great Greek physicist Archimedes (287–212 BCE).

In the following centuries, Aristarchus's ideas were not widely appreciated, in part owing to the influence of the Aristotle, who had common sense on his side. If the earth is moving, why do we not feel that it is moving? Why do we not fall off? The Greeks at the time of Aristotle had no answers to these questions because they did not understand the ideas of inertia and of gravity. These concepts were understood only after the sixteenth century and will be discussed in later chapters.

After Aristarchus, the greatest Greek astronomer was Hipparchus, who lived in the second century BCE. Using trigonometry, he refined Aristarchus's work, obtaining the relative sizes of the earth, sun, and moon more accurately. Nevertheless, Hipparchus accepted Aristotle's view that the sun moves around the earth, rather than the reverse.

The reason for Hipparchus's conclusion is that if the earth moves around the sun, then the stars should appear to shift their positions as seen from the earth. This shifting is known as "stellar parallax," and Hipparchus did not observe it. To understand parallax, hold a finger a foot in front of your face, and look at the finger, alternately with one eye closed and then with the other eye closed. The finger appears to be in different positions relative to the background in the two cases. The reason is that the two eyes, being separated, view the finger from different angles. The angular separation of the two apparent positions of the finger is known as parallax. Knowing the separation of the eyes, one can calculate the distance of the finger from the difference in observed angle. The farther away an object is from us, the smaller the angular separation when we view it from different positions. We are not confined to the diffeernt angles seen by our two eyes. For far-off objects, we can move to different places on earth, and see how the angle changes. Then we can calculate the distance of the object by how much the angle differs when viewed from the two places.

If the earth moves around the sun, then the positions of stars should appear different in winter and summer, when the earth is at opposite sides of its orbit. The difference in observed position relative to more distant stars is called "stellar parallax." We illustrate steller parallax in Figure 1.2.

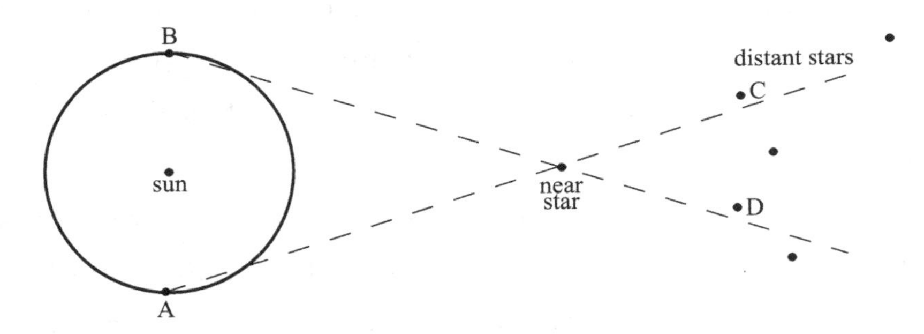

Figure 1.2: Stellar parallax. When the earth is at A in its orbit, a nearby star seems near distant star C, but when the earth is at B, the nearby star seems near distant star D. So the nearby star apparently moves with respect to distant stars. The figure is not drawn to scale.

But Hipparchus did not observe stellar parallax and so concluded that the earth did not move. Hipparchus's idea to look for stellar parallax was a good one, and with telescopes we can observe stellar parallax for the closer stars. But the amount of parallax decreases with distance of the observed object, and even the closest stars are too far away for Hipparchus to have observed parallax with the measuring instrument he had at his disposal: namely, his eyes. So Hipparchus missed the heliocentric view of the sun and planets because he could not believe that stars were so far away from us.

The astronomer Claudius Ptolomy of Alexandria, who flourished in the second century CE, codified the prevailing ideas of his time and added ideas of his own as a result of his observations. According to the Ptolomaic system, the earth is a stationary globe at the center of the universe. The basic motions of the sun, moon, stars, and planets are circles around the earth. The stars are fixed on a "celestial sphere," which rotates once every twenty-four hours. The planets could not be stationary on the same celestial sphere because they move with respect to the stars.

In order for the picture to agree with careful observations, Ptolemy had to assume that the planets move in small circles, called "epicycles," whose centers move in large circles around the earth. Epicycles were first proposed some four hundred years before Ptolemy, when it was realized by observation that the planets do not move in Aristotelian circles. However, the circle was supposed to be a "perfect" curve, and therefore, the description of the motion of the planets was described by circles within circles.

After the Ptolomaic system was explained to King Alfonso X of Castile and Leon (1221–1284), the king is supposed to have said, "If the Lord Almighty had consulted me before embarking upon Creation, I should have recommended something simpler."

1.4 The Copernican revolution

Despite the skepticism of King Alfonso and others, the Ptolomaic system was widely believed until the Polish astronomer, Nicolaus Copernicus (1473–1543) revived the Aristarchan heliocentric view of the sun and planets. All the planets, according to Copernicus, revolve around the sun. He boldly explained the absence of stellar parallax by assuming that the stars were too distant for parallax to be observed. Copernicus completed his treatise around 1530, but it was not published until just before he died. It appeared under the title, *De Revolutionibus Orbium Coelestium (On the Revolution of the Celestial Sphere)*. The title, which seems to contradict the main ideas of the book, was the choice of the publisher.

Because Copernicus assumed that the planets go around the sun in circles (their paths are actually approximate ellipses), he also had to postulate the existence of epicycles. In fact, the Copernican model was almost as complicated as the Ptolemaian model, and the predictions were not much more accurate. However, Copernicus was able to calculate the relative speeds and distances of the planets from the sun, something that did not seem possible in the earch-centered model.

We speak of the "Copernican revolution" because his work began a profound change in the way we look upon the universe. Copernicus shifted our point of view from the earth to the sun, but later it was realized that the sun is just one star in a vast collection of stars known as the Milky Way. Still later it was understood that the Milky was is just one huge collection of stars (called a galaxy) out of many.

For complicated reasons (which we do not care to go into), the Catholic Church had adopted the Aristotelian view that the earth is the center of the universe. At the time of Copernicus, the Church took the view that not only religion but science came under its purview, and it tried to supress the ideas of Copernicus. Despite these attempts, Copernicus's work survived and was the first in a series of great scientific revolutions that deeply changed the way we think about the unvierse.

Chapter 2

The Solar System and Beyond

> Philosophy is written in a great book which is always open in front of our eyes (I mean: the universe), but one cannot understand it without first applying himself to understanding its language and knowing the characters in which it is written. It is written in the mathematical language.
>
> —Galileo Galilei (1564–1642)

2.1 Elliptical orbits

The Copernican picture of the sun and planets (the solar system) was a great advance over the Ptolomaic picture, but the Copernican idea suffered from the false assumption that the planets go around the sun in circles. About a century later the German astronomer and mathematician, Johannes Kepler (1571–1630), took the next big step. Using records of the precise observations of the planets by the Danish astronomer Tycho Brahe (1546–1601), Kepler concluded that the planets do not go around the sun in circles but rather in oval-shaped orbits called ellipses.

In the interior of an ellipse are two points, called foci (the singular is focus), on opposite sides of the center. The sum of the distances from the curve of the ellipse to the two foci is a constant. The closer together the two foci are relative to the size of the ellipse, the closer the ellipse is to a circle. In the limit that the two foci coincide, the curve is a circle. The orbits of the planets known at the time of Kepler are very nearly circles. A drawing of an ellipse is given in Figure 2.1. The ellipse is very elongated compared to the orbits of the planets.

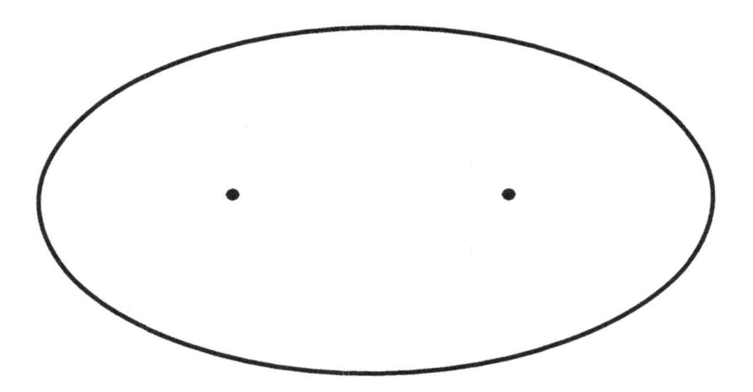

Figure 2.1: An ellipse, showing the two foci.

Kepler concluded from Brahe's observations that the sun is not located at the center of the ellipse but at one of the two foci. Nothing at all is at the other focus. The statement that the planets move in ellipses around the sun with the sun at one focus is known as Kepler's first law. From Tycho Brahe's observations, Kepler deduced two other laws of planetary motion.

Kepler's second law says that as a planet moves around the sun, an imaginary line from the sun to the planet sweeps out equal areas in equal times. As a consequence, the planet moves fastest when it is closest to the sun (the "perihelion") and moves slowest when it is farthest from the sun (the "aphelion"). A drawing illustrating Kepler's second law is given in Figure 2.2. The two shaded areas shown in the figure are equal, so that it takes the same amount of time for a planet to go from A to B as it takes for the planet to go from C to D. The actual planetary orbits are more nearly circular than the drawing shows.

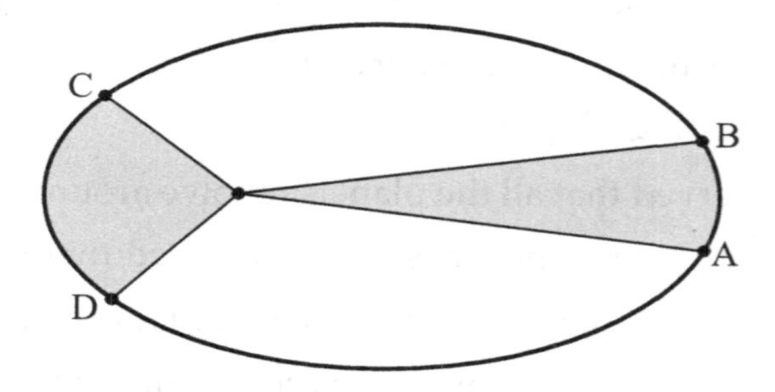

Figure 2.2: Illustration of Kepler's second law.

The third law says that the square of the period of a planet (the period is the time it takes the planet to make one complete orbit) is proportional to the cube of the semimajor axis (the line from the center of the ellipse through one focus to the curve of the ellipse). (The square of a number is the number multiplied by itself, while the cube of a number is the number multiplied twice by itself.) Kepler's third law makes quantitative the observed fact that the larger the orbit of a planet, the slower it moves.

Kepler's three laws are very accurate, but we now know that they are not exact. For example, a very slight discrepancy in the orbit of Mercury was discovered in the nineteenth century, showing that the orbit is not exactly an ellipse. Part of the discrepancy in Mercury's orbit could be attributed to perturbations from the gravity of other planets, but a residual deviation remained. This deviation was not explained until the general theory of relativity was proposed in the early twentieth century, discussed in Chapter 10.

While Kepler was doing the work that led to his three laws of planetary motion, he decided that the reason the planets move around the sun is that the sun exerts some influence over the planets. This idea was a precurser of the theory of gravitation of Isaac Newton (1642–1727), an English physicist.

Kepler observed that all the planets revolve around the sun in the same direction. In ancient times it was believed that the planets revolve around the earth in the same direction. However, at times during their orbits, the planets are observed to undergo what is known as retrograde motion—that is, they temporarily reverse direction as observed from earth. Retrograde motion was originally explained in terms of epicycles, but Kepler understood the motion from his three laws. A planet farther from the sun than the earth moves more slowly than the earth. When the earth and the planet are moving on the same side of the sun, because the earth is moving faster, the planet moves in the opposite direction *relative to the earth*. Of course, when planets are moving on the opposite side of the sun, they are moving in a direction opposite to that of the earth.

Our present understanding of why the planets revolve around the sun in the same direction is that the solar system evolved from a rotating cloud of gas. As the cloud condensed into the sun and the planets, the rotary motion persisted. That is why the sun spins around its axis in the same direction as the planets revolve around the sun. The sun and the planets are said to have "angular momentum," a concept discussed in Chapter 7.

2.2 Galileo's contributions

Galileo Galilei (1564–1642), an Italian, was a contemporary of Kepler and the first great "modern" physicist. (At that time physics was called philosophy.) Galileo made revolutionary changes in our ideas of motion as well as our ideas about astronomy. He and Kepler both observed the supernova of 1604, an object that suddenly brightened in the sky. From the absence of parallax, Galileo realized that the object was at stellar distances and therefore exceedingly bright. We now know that a supernova is an exploding star. We shall have more to say about supernovae in Chapter 18.

Galileo did not invent the telescope but as soon as he heard about it he built one of his own and was the first person to use it to look at the sky. One of his most significant discoveries was the first observation of four of the moons of Jupiter. By observing these satellites at different times, Galileo established that they were orbiting Jupiter and not the earth. This was the first instance to be established that heavenly bodies orbited an object other than the earth. Of course, Kepler's work showed it was much simpler to assume that the planets orbit the sun rather than the earth, but many astronomers still held the earth-centered point of view.

Galileo also studied sunspots, which are relatively dark spots on the surface of the sun. Before Galileo, astronomers believed that the spots were not on the sun but moved in front of it. Galileo, however, noted that the spots apparently move slower when near the edge of the sun than when near the center. He explained this behavior by assuming that the sun rotates at constant angular speed. When a spot is near the edge of the sun the spot moves mostly toward the earth or away from it as the sun rotates, making the spot's sideways motion appear slower. Galileo's observation scandalized those astronomers who believed the sun must be unblemished and must not rotate.

In his great work, *Dialogue on the Two Chief World Systems*, Galileo argued for the heliocentric picture of the sun and planets. He did not directly advocate the Copernican picture because it was forbidden by the church. Instead, he introduced three characters: Salviati, Simplico, and Sagredo. Salviati gave arguments for the heliocentric view, based on Galileo's observations. Simplico argued for the Aristotelian view, and Sagredo listened to both Salviati and Simplico but was persuaded by the superior arguments of Salviati. Despite Galileo's saying in his preface that he did not personally believe in the Copernican system, the Church banned the book (as it had banned the works of Copernicus and Kepler) and put Galileo under house arrest for the rest of his life.

2.3 The stars

The Copernican revolution was the beginning of a profound change in the way we look at the universe. Copernicus's work led to the abandonment of the idea that the earth is the center of the universe, to be replaced by the newer idea that it is the sun. Later astronomers came to the conclusion that the sun is not at the center of the universe but is only one star in a huge collection of perhaps about two hundred billion stars known as the Milky Way galaxy. Still later, other astronomers discovered that the Milky Way is only one of many billions of galaxies in our visible universe. It has been estimated that as many as a trillion (1000 billion) galaxies are in the visible universe.

Copernicus knew that the stars are not part of the solar system. The stars apparently rotate around the earth once every twenty-four hours, but the reason for this apparent rotation is that the earth rotates once a day around an axis that goes through the north and south poles of the earth and through its center. The stars seemed to Copernicus and Galileo to be fixed in space. We now know that the stars also move, but because they are so far away they appear to be stationary in space.

Let us discuss the reason why we do not see stars move, except for their apparent motion caused by the rotation of the earth. If you look at at stationary car in the distance, light from the front and back ends will enter your eyes at different angles. The difference in angles gives you the perception that the car has a length. If the car is twice as far away, the angle is only half the size (to a good approximation), and so the apparent length is only half as much. Our brains compensate for this decrease in apparent length and so we realize that the car has not changed size. With a star, which looks very much like a point of light, our brains have no way of compensating, so we do not have any idea how far away a star is just by looking at it. Now imagine a person walking from one end of the car to the other end in a certain time. If the car and the person are twice as far away, the person apparently walks half the distance (because the car seems half as long) in the same amount of time, and so seems to move only half as fast. The farther away a person is, the more slowly he seems to walk. This same effect holds for other moving things as well, so, for example, an airplane high in the sky seems to move more slowly than an airplane close to the ground. The stars are so far away that their apparent motion is too slow for us to observe. As we shall see in Chapter 9, we can deduce some information about the speed of stars by observing how the frequency of the light emitted by the stars changes because of their motion. (The effect is called the Doppler effect.)

The information astronomers obtain from observation must be interpreted in terms of physical laws of nature, or, more accurately, in terms of our theories about the way nature behaves. It is theory that tells us, for example, why stars shine, as we shall discuss in Chapter 15.

Chapter 3

Newton's Ideas about Space and Time

God in his wisdom made the fly, and then forgot to tell us why.

—Ogden Nash (1902-1971)

In order to discuss the universe in a useful way, we need to know something about the forces of nature. It is also essential to have some knowledge about the concepts of space, time, and motion. We devote a number of chapters to these topics. We begin with ideas about spece and time that are useful to describe ordinary happenings on earth but that need to be modified if we are to understand nature at a deeper level. In subsequent chapters we describe other topics that will help us understand the universe.

3.1 Infinite space

For more than 200 years, Isaac Newton's ideas of space and time were generally accepted by the physics community, but were overthrown in the 20th century by the theories of relativity and quantum mechanics. The special and general theories of relativity were by Albert Einstein (1879–1955), a German-born physicist who fled Germany in 1933 when Hitler came to power, and who then became an American. The ideas of quantum physics began in 1900, the last year of the 19th century, with the work of Max Planck, and developed over the years by many physicists, culminating in the 1920s with the invention of quantum mechanics.

Newton's concept of space is that it exists independently of matter, and stretches out to infinity in three mutually perpendicular directions. The number of mutually perpendicular lines one can draw through a point is called the *dimension* of the space. To our eyes, space is three-dimensional, although there have been speculations that space contains additional dimensions that we are unaware of.

We can never measure an infinite distance, or measure anything infinite directly. So the statement that space is infinite is a theoretical idea that can never be tested. However, we may be able to test certain consequences of the idea that space is infinite. If the consequences are found to be correct (within the errors of our measurements), we gain confidence in the soundness of the idea. But this confidence is always subject to revision, for it is possible that a future measurement may show a contradiction to a prediction of the theory.

Newton's space has additional properties. First of all, it is "flat," which means that it obeys Euclidean geometry. With such a geometry, (straight) parallel lines never meet (or, what essentially amounts to the same thing, parallel lines meet only at infinity). In Euclidean geometry, this statement about parallel lines is sometimes taken as a definition of what parallel means. In flat or Euclidean geometry, only one line can be drawn through a point external to a given line which is parallel to the given line. This statement (taken together with the definition of parallel lines) is equivalent to one of the axioms on which Euclidean geometry rests. The statement cannot be proved from the other axioms of Euclidean geometry. In Figure 3.1 we illustrate parallel lines in Euclidean geometry.

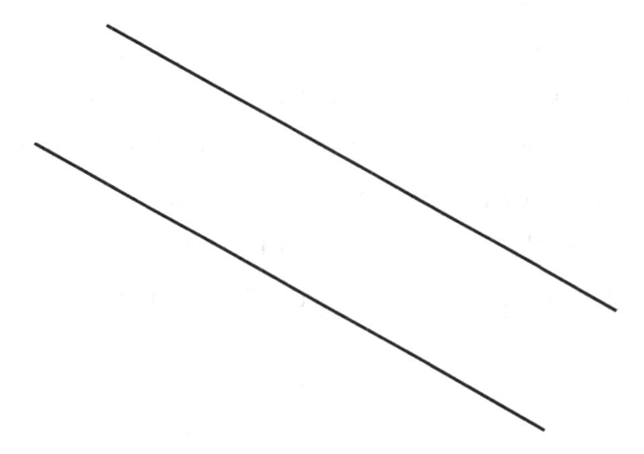

Figure 3.1: Parallel lines in Euclidean geometry. The lines may be extended (in principle) indefinitely in both directions without intersecting.

In later mathematical developments, it became clear that the definition and/or the axiom could be given up and the resulting "non-Euclidean" geometry could still be consistent. An axiom is something that is supposed to be "self-evident." With these developments, it is more appropriate to call the statements that Euclid accepted without proof "postulates." Postulates are initial assumptions for a mathematical system, and do not need to be self-evident or true in nature. Instead of calling the statement about parallel lines not intersecting a definition, we may call it a postulate. Then another definition has to be given of what makes two lines parallel. In the case of non-Euclidean geometry, the definition is complicated, and to give such a definition would take us too far afield.

A "straight line" between two points on a curved surface is defined to be the shortest distance between those two points on the surface. As an example of a non-Euclidean geometry, consider the surface of a sphere, which is curved. A "straight line" on the surface of a sphere is a *great circle,* which is a circle whose center is at the center of the sphere. On a sphere, parallel lines do intersect. For example, the lines of longitude on the surface of the earth are great circles. They are all parallel locally at the equator and all intersect at the north and south poles. In Figure 3.2 we illustrate great circles.

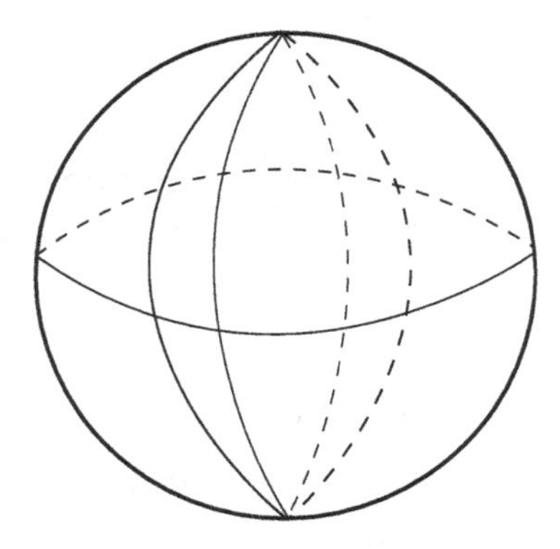

Figure 3.2: Great circles on a sphere.

On the other hand, the lines of latitude do not intersect, but, except for the equator itself, are not great circles. Therefore, the lines of latitude are considered to be curved. Because parallel lines mean parallel "straight" lines, the lines of latitude are not considered to be parallel.

We must distinguish a flat three-dimensional space from a flat space in two dimensions. Imagine beings confined to live on a plane and unaware of a third dimension. Actually, life as we know it cannot exist in two dimensions but needs a third. We can see this as follows: In three dimensions we take in food in a cavity which runs through the body in a continuous fashion so that we can eliminate waste at at the other end of the cavity. But in two dimensions a cavity running through the body and open to the outside in two places divides the individual into two distinct separated parts. This is unlike the situation in three dimensions, in which a long hole can run through a body while the body remains connected. Of course, in principle a long hole can exist in a two-dimensional body, but it can have only one opening to the outside so that the waste must be eliminated through the same opening in which food is taken in.

Ignoring the problem of how life can exist in two dimensions (we believe it can't), let us examine the perception of a two-dimensional being observing a sphere passing through his plane. The sphere looks like a point as it just touches the plane, then becomes a small circle which grows to a maximum size, shrinks again to a point, and then disappears. The two-dimensional being observing the sphere from the outside will not see a point become a circle but will see it become a line, grow, and then shrink again until it disappears.

If a circle rotates, it looks to the viewer as a line that doesn't change in size. We say that the circle has a symmetry that leaves it looking invariant under any rotation between zero and 360 degrees.

No other two-dimensional curve has as much symmetry. For example, a square has a symmetry through a rotation of 90 degrees and multiples thereof, but under a rotation of less than 90 degrees, the length of the externally observed line will change.

We in three dimensions cannot really see a whole sphere, but it appears to us only as a disk. It is our depth perception that gives our brains information that allows us to interpret a disk as a sphere, and depth perception does not work with far-off objects like the moon. We have other evidence that the moon is a sphere rather than a disk. The most direct evidence comes from artificial satellites sent to the moon that revolve around the moon.

A sphere has more symmetry than any other shaped object because it doesn't change its apparent shape no matter how it rotates. Of course, the bright part of the moon does change its shape because the bright part is the part illuminated by the sun that is observable on earth, and that part changes as the moon goes around the earth. But sometimes the part of the moon in shadow can be faintly observed from the earth, and then it is seen that the bright part plus the part in shadow appear to make a full disk.

If a four-dimensional analogy of a sphere passed through our space of three dimensions, we would perceive it as a point becoming an apparent disk which grows to a maximum, then shrinks, and disappears. With depth perception, we could observe in principle that the disk was really a sphere.

We return to three-dimensional space. According to Newton, empty space (called a vacuum) exists independently of any matter that happens to be located within it. (How empty space could be observed is a question that Newton could not answer.) In any case, according to Newton, space has the property that it can transmit the gravitational force between distant objects. However, empty space cannot transmit sound, which needs a medium, such as air, in which to travel. Newton was uncomfortable with this so-called gravitational "action at a distance," but he had no other satisfactory explanation for how gravity worked. Implicit in Newton's idea of action at a distance is that the gravitational force between separated bodies acts instantaneously, no matter how far apart the bodies are from each other. This idea, which implies that the speed of action of the gravitational force is infinite, has turned out to be wrong, as we shall discuss in later chapters. Nevertheless, in many commonly occuring situations, the idea is a sufficiently good approximation to be very useful. That is why we still use Newton's law of gravity.

3.2 Infinite time

According to Newton, time is also infinite, but it differs from space in several crucial respects. Time is one dimensional, extending from the infinite past to the infinite future. Although we can move at will in any direction in space, moving up or down, forward or backward, or side to side, in time we move inexorably forward. Furthermore, Newton believed that we all move forward at the same rate, helpless to affect our passage through time. These ideas of Newton also turned out to be wrong, but a very useful approximation in most ordinary situations.

Some, but not all, of the equations of modern physics look the same whether time moves forward or backward. For example, if time were to go backward, the earth would spin in the opposite direction and would also go around the sun in the opposite direction. A distant observer of the earth and the sun would not be able to tell by these motions whether time was going forward or backward.

However, most observations enable us to tell that time is going forward. Consider a film of an egg falling to the floor and breaking. Run backwards, such a film would show the broken egg put itself together and rise to the table from which it fell. Anyone familiar with the way things happen in real life would know that the film was running backwards and that such an event would not happen in reality.

The ways in which Newton's ideas about space and time go wrong are subtle and not at all apparent in our daily activities on earth. Newton was one of the greatest scientists and mathematicians in history, and he would not have made obvious blunders about the nature of space and time. His subtle errors arose primarily because he did not have at his disposal modern precise experiments, which show deviations from his assumptions. He also could not know of the future invention of electromagnetic theory in the nineteenth century by James Clerk Maxwell (1831–79), which had implications (discovered by Einstein early in the twentieth century) that contradicted Newton's ideas. We discuss these developments in later chapters.

3.3 Scalars and vectors

It is obvious that we can discuss space and time only because we exist. Obviously many things exist in space. Newton discussed the properties of space and time in empty space because he assumed that the presence of things in space did not influence those properties. As we have remarked, he was not strictly correct in his assumption. But for the present, we shall ignore how space and time are affected by matter and motion, and shall discuss objects moving in Newton's space and time. It is clear that motion depends on time as well as space because motion in space takes place during time. To discuss motion we need to make precise the familiar ideas of distance, velocity, and acceleration.

But first we need to introduce the ideas of scalars and vectors. If something we measure can be specified by a single number (a magnitude) along with its appropriate unit, that something is called a *scalar*. The temperature of a glass of water is a scalar because it is specified by a magnitude, for example, 20 and a unit, for example, degrees Celsius, abbreviated ° C). In other units, the same temperature will have a different magnitude, for example, 20° C is the same as 68 degrees Fahrenheit (68° F).

However, something may require more than one number for its specification. For example, if we say that Boston is about 330 kilometers (a little over 200 miles) from New York, we have not supplied enough information to a traveler who wants to go from New York to Boston. We must also specify the direction the traveler has to go, in this case, approximately northeast. The direction one has to go may also be specified by a number, which may be chosen to be the angle that the direction makes with north. In this example, only one angle is sufficient to specify the direction because the motion is confined to the surface of the earth. If we were talking about the distance to a bird in a tree, we would need a second angle, which would specify the angle made with the horizontal. A quantity that has both *magnitude* and *direction* is called a *vector*. The magnitude of a vector is sometimes called a scalar.

The *distance* between two points is a scalar. The distance between New York and Washington, DC, is about the same as the distance between New York and Boston. But to get to Washington from New York, a traveler must go in almost the opposite direction as to get to Boston. Physicists use the technical term *displacement* for the vector which gives both the distance and direction from one point to another.

Scientists often use a shorthand notation by denoting frequently-used quantities by symbols rather than words. The symbol for a quantity is often a single letter. Use of symbols instead of words to denote frequently-used concepts not only saves time but also often aids in thinking about the subject. Also, when scientists use symbols, they are better able to employ the powerful tools of mathematics to manipulate the symbols in order to shed additional light on the subject.

We usually denote a scalar by a single italic letter and a vector by a letter in bold face. We often denote the distance between two points by the letter s and displacement by the letter $\mathbf{s}$. The quantity s is the magnitude of the vector $\mathbf{s}$. Sometimes we have to deal with more than one distance (or other quantity) in the same problem, and then we denote the different distances either by different letters or by the same letter with different subscripts. It does not matter what symbol we use for any quantity so long as we define it so that we know what we are talking about.

In discussing the distance between two places, say Boston and New York, we have to distinguish between two different quantities. The first is the straight-line distance (as the crow flies), which is also the magnitude of the displacement. The second quantity is the distance we might travel if we went by car on curved roads. When we use the word distance without any qualification, we mean the straight-line distance unless it is clear from the context that we mean some other distance. If you walk in a circle until you return to your starting place, the distance you have walked is obviously not zero. However, your (vector) displacement is zero, and therefore the magnitude of your displacement is also zero. If a vector has magnitude zero, its direction is not defined.

When we talk about the straight-line distance between Boston and New York, we gloss over an important fact. A really straight line between those two cities will take us through the interior of the earth. (This is even more obvious of a straight-line path between New York and Tokyo.) As we have said, what we mean by a straight line distance on a curved surface is the shortest distance between two points on the surface in question. If we idealize the earth as a smooth sphere (without mountains and without flattening at the poles), then the shortest distance between two points on the earth's surface is part of a great circle containing the two points. Given any two points on the earth's surface, there is only one great circle through them except if the points are half way around the world from each other. If the points are not half way around the world, then, starting from one of the points, there are two ways one can get to the other point along a great circle. Either of these paths is considered a "straight line" on the sphere, and the shorter of the two paths is the shortest distance between the two points. If the points are half way around the world, then there are infinitely many great circles through the points, and it is the same distance between the points on any of those great circles. Any path on the surface between two points which is not along a great circle is considered "curved."

As an example, Los Angeles and Atlanta are both approximately at the same latitude of 34° north of the equator. (Actually, Atlanta is slightly south of Los Angeles by less than a degree of latitude.) However, if an airplane wanted to fly from Los Angeles to Atlanta by the shortest route, it would not fly along 34° latitude but along a great circle connecting the two cities, which would take the airplane *north* of 34° latitude for all of its route except for the two ends.

In many problems, the curvature of the surface in question is so small that it does not need to be considered. When this is the case, we shall not mention curvature, and the surface will be considered to be flat.

Let us return to vectors. We can describe a vector in a an alternative way to giving its magnitude and direction by giving numbers called the *components* of the vector. For example, suppose you see a woodpecker on a pole. You can describe the displacement of the bird from your own position by giving two numbers to locate the position of the pole and then a third number to locate the position of the bird on the pole. To reach the pole you might have to walk 30 meters north and then 40 meters west. Then, to reach the bird, you might have to climb 10 meters up the pole. (A meter is a little longer than three feet.) To describe the vector displacement, we use three numbers, each with its unit, which in our example is the meter, abbreviated m: 30 m north, 40 m west, 10 m up. These three numbers are called the *components* of the vector displacement.

Specifying the components of the displacement serves as a useful alternative to specifying its magnitude and direction. However, if we want to use the magnitude and direction, we still must specify three numbers. This is because in general *two* angles are necessary to specify the direction, for example, the angle with respect to north and the angle above the horizontal. The magnitude of the vector is the third number required for its specification.

In order to make a measurement of the displacement, it is useful to imagine three mutually perpendicular lines, called "axes," which emerge from any convenient place, called the "origin." The axes may be called the x, y, and z axes. The directions of the lines are any convenient perpendicular directions. For example, the x axis might be east, the y axis north, and the z axis up. We can measure the displacement, or any other vector in space, by measuring its components along the three axes.

Consider a vector, which we call $\mathbf{F}$. We have in mind that $\mathbf{F}$ stands for the force on an object, which is loosely defined as a push or pull on the object. A force not only has a magnitude but a direction in which it acts, and so it meets the definition of a vector. (We could equally well have considered an arbitrary vector and given it another name.) Then we denote its components in the x, y, and z directions by F_x, F_y, and F_z respectively.

We can picture a vector by a straight line segment with an arrowhead at one end of it. The length of the line segment stands for the magnitude of the vector, and the direction the arrow points stands for the direction of the vector. The end of the line segment with the arrowhead is called the "head" of the vector, and the other end is called its "tail."

In Figure 3.3 we show a vector **F** with its three components along an arbitrary set of mutually perpendicular axes. We cannot accurately portray three dimensions on a two-dimensional sheet of paper, so the y axis should be imagined as coming out of the paper.

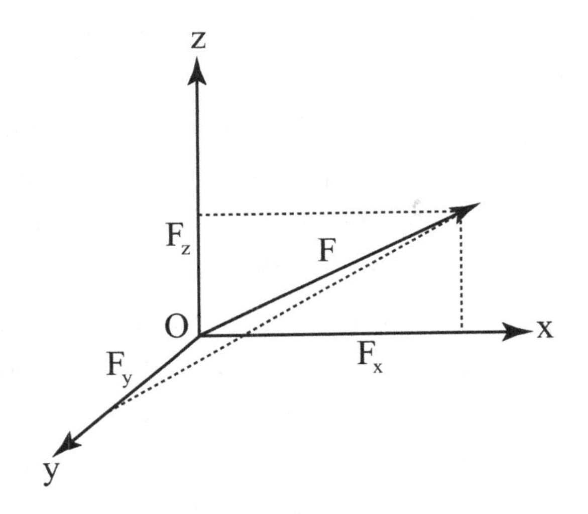

Figure 3.3: A vector **F** with its components along three perpendicular axes. The y axis should be considered as coming out of the paper.

We need to discuss many quantities in physics, but there are only a limited number of letters in the alphabet. Sometimes we also use the Greek alphabet and certain special symbols as well. Also, sometimes we use the same letter to denote different quantities. For example, the letter s (in italic) is often used for distance and s (not italic) for second, a unit of time. If it is not clear from context, then a different letter should be used for each different quantity.

3.4 Four-dimensional spacetime

It is no accident that three numbers are required to specify a vector, such as displacement, independently of whether we use its magnitude and direction or its components. This fact arises because we live in a world of three spatial dimensions. More accurately, we perceive a world of three spatial dimensions. There might exist additional dimensions that we have not yet been able to perceive (see Chapter 20). Because space appears to have three dimensions, displacement is often called a *3-vector*.

We can easily picture a displacement on a flat surface (a plane), which requires only two numbers for its specification, or displacement along a straight line, which requires only one number. On the other hand, we find it hard to imagine what a world with more than three spatial dimensions would be like. Such a world is outside our experience. Mathematicians, however, have no trouble in using mathematical symbols to "describe" worlds with any number of spatial dimensions, even though those mathematicians cannot visualize more than three dimensions any better than the rest of us.

Many people have wondered why our space has three dimensions. Some people have even speculated about reasons for this fact, but nobody has yet come up with a satisfactory answer. Some people have speculated that there are additional dimensions, but that these are in directions in which space is curved, and that the extra dimensions curl around in such a small distance that we are unaware of them. However, at present no one has demonstrated the existence of extra spatial dimensions in our universe.

In physics the concept of time seems just as basic as the concept of space. Time is commonly thought of as a scalar, because only one number is required for its specification. For example, a family may sit down to dinner at 7 p.m. This time is understood because we have standard units of time (hours, minutes, seconds, to name a few), and standard time zones. Just as we have standard rulers to measure distance, we have standard clocks to measure time.

But there is a sense in which time is not a scalar. Suppose you are asked to describe *where* and *when* you saw a woodpecker on a pole. You might respond by giving four numbers: 30 meters north, 40 meters west, 10 meters up, 4:20 p.m. These numbers might be considered the components of a 4-vector. It is not the usual vector, to be sure, because three components specify distances and one component specifies a time. Thus, 4-vectors are not vectors in ordinary space but vectors in spacetime. It is in this sense that time is sometimes called the fourth dimension. There is an apparent difficulty with this interpretation, because time has different units than distance. However, because the product of speed and time has the same units as distance, we can overcome the problem of units by multiplying time by a speed, usually the speed of light. Then the fourth number is the distance traveled by a beam of light in the time in question.

Einstein's special and general theories of relativity (discussed in

Chapter 10) are conveniently described in terms of 4-dimensional spacetime. In these theories ordinary distances in space and ordinary time intervals are not true scalars, but the magnitude of the spacetime displacement is a scalar. The magnitude of the spacetime displacement depends on the properties of the spacetime, and these properties are determined by what is called a "metric." However, for many purposes the approximation that distance and time are separately scalars is a good one. In fact, the approximation is good whenever Newtonian mechanics is a good approximation, and Newtonian mechanics provides an excellent description of the motion of ordinary macroscopic objects on the earth. We discuss Newton's laws of motion in Chapter 5 and his theory of gravity in Chapter 6. So, until we discuss relativity, we shall treat distance and time separately as scalars.

3.5 Velocity and acceleration

A number of scalars and vectors are important in the study of motion. We have already discussed several scalars (temperature, distance, and time) and vectors (displacement and force). We next consider some of these quantities in more detail and introduce some others.

Suppose a traveler can drive to Boston from New York, a distance, say, of 350 kilometers (km) in 5 hours (hr). Then, knowing both the distance and the time, we can define a derived quantity, the *average speed*, which is the distance traveled divided by the time required to go that distance. In our example, the average speed is 70 kilometers per hour (70 km/hr), which is obtained by dividing 350 km by 5 hr. A speed of 70 km/hr is about 44 miles per hour (mi/hr).

Let us introduce the symbol v for "speed" and $\bar{v}$ for "average speed." (Our notation here is that a line on top of a symbol denotes its average value. In later chapters we use the line with a different meaning.) Our *definition* of average speed of a body is that it is the distance traveled, which we denote by the symbol s, divided by the elapsed time, which we denote by the symbol t. Thus, using our symbols, we can write $\bar{v} = s/t$. By using symbols, we not only shorten our notation but can manipulate the symbols algebraically. However, we shall rarely use algebra in this book. Unfortunately, there is a price to pay for using symbols: we either have to remember what the symbols mean or write down their definitions.

If our driver looks at his speedometer from time to time during the trip from New York to Boston, he will notice that the speedometer does not always read 70 km/hr, but sometimes more and sometimes less. In order to know how fast the car is going at any given time, we need the concept of speed at a given instant of time. But an instant of time is no time at all. If speed is distance divided by time, and the time is zero, how can we define the speed? We learned quite early in arithmetic that we are not allowed to divide by zero.

We obtain a solution to the problem as follows: We divide the total distance into a number of smaller distance segments, and measure the time interval required to travel each segment. Then the average speed during each segment can be calculated by dividing the segment by the time interval to traverse it. As we make the distance segments smaller and smaller, the corresponding time intervals to traverse each of them will also become smaller. When the distance segments and time intervals are so small that the car cannot appreciably change its speed during a single interval, then the average speed in the interval is approximately the same as the speed throughout the interval. Thus, we can define the *speed* at a given time as the average speed during a very small time interval which includes the given time. A more precise definition is that the speed is the limit of the average speed as both the distance segment and time interval become arbitrarily small. The branch of mathematics that is suited to calculate with quantities that become arbitrarily small is called the calculus. We do not discuss the calculus in detail in this book.

Just as we define the scalar speed as a distance segment divided by the corresponding time interval, so we can define a vector *velocity* as a displacement segment divided by the corresponding time interval. The vector velocity is often denoted by the symbol $\mathbf{v}$. The scalar speed v is the magnitude of the velocity. Unfortunately, not even physicists are always precise in their language, and sometimes use the word "velocity" when they mean "speed." Because of this imprecision, one is often forced to infer the precise meaning from context.

We previously mentioned a car changing its speed as it travels along. The car also changes its direction from time to time. Whenever a body changes its speed, its direction, or both, it is changing its velocity. If the velocity of an object changes, we say that it undergoes *acceleration*, usually denoted by the symbol $\mathbf{a}$. Like velocity, acceleration is a vector, although the magnitude of the acceleration, a scalar, is also called acceleration. Because we use the same word for vector and scalar acceleration, we have to make the meaning clear by context. The symbols, however, are different, being **a** for the vector and a for the scalar.

We can define the acceleration as the change in velocity during a very short time interval divided by the time interval, or, in other words, acceleration is the rate of change of velocity. In the case of a car speeding up or slowing down on a straight road, the car accelerates by virtue of a change in speed. In the case of a car going around a curve at constant speed, the car accelerates by virtue of a change in direction. Of course, a car can change its speed while going around a curve, and this too is acceleration. In ordinary language, slowing down is sometimes called "deceleration," but we often use the same word acceleration for slowing down, speeding up, or changing direction with or without a change in speed.

Many people confuse the concepts of velocity and acceleration. Because velocity and acceleration are vectors, each has a direction, and these directions may be different. If the velocity and acceleration of an object are in the same direction, the object moves in a straight line and speeds up. If the acceleration is in the opposite direction to the velocity, the object moves in a straight line and slows down. If the direction of the acceleration makes an angle with the direction of the velocity, the object moves in a curved path.

It is possible for the velocity of an object to be zero at a certain instant, while at that same instant the acceleration is not zero. For example, if you throw a ball straight up in the air, then, when the ball is at its highest point above the ground, the velocity of the ball is zero. Its acceleration at that point is not zero, however, because even though the velocity is momentarily zero, the velocity is changing. If the velocity and acceleration were both zero, the velocity would not change, and the ball would remain motionless in the air, clearly in conflict with observation.

Velocity and acceleration have different units. Because velocity is a displacement divided by a time, the unit of velocity is a unit of displacement, say, meters (m) divided by a unit of time, say, seconds (s), or m/s. The unit of acceleration is the unit of velocity (m/s) divided by the unit of time (s), or m/s/s. This unit is also written m/s^2.

Chapter 4

Early Ideas of Motion

> Paradox. An assertion that is essentially self-contradictory, although based on a valid deduction from acceptable premises.
>
> —*The American Heritage Dictionary* (Houghton Mifflin, 1985)

4.1 Zeno's paradoxes

We interrupt our treatment of the space and time in order to discuss some ideas about motion. The study of motion is a necessary prerequisite to understanding motion in the universe. We begin with early ideas about motion.

The discovery of an apparent paradox is often the first step toward making a fundamental advance in logic or science. If our premises lead to a contradiction, they are inherently inconsistent. In mathematics, we ask only for a consistent set of premises or hypotheses, no matter how implausible they may seem. A strange set of hypotheses may lead to a strange mathematics, but that's OK.

However, if we make a set of mathematical hypotheses and arrive at a contradiction, we have to examine our set of hypotheses in order to discard or replace members of the set until we arrive at a consistent scheme. This can sometimes be done in a variety of different ways, and each way may lead to a different mathematical system. How do we choose between the various schemes? The answer is that we don't. Each consistent scheme is equally valid, although some may be richer or more beautiful than others.

In science, on the other hand, we want our ideas to reflect nature as closely as possible. If we arrive at a paradox in science, at least one of our premises must be *wrong* because nature's laws must be consistent. The existence of the universe is the proof of its consistency. So if we encounter a paradox in science, we conclude that, in some subtle way, nature is not behaving as we had assumed. It is often an exciting quest to find the wrong premise and to discover what must replace it.

The history of physics contains a number of instances in which the discovery of a paradox and its resolution has led to a crucial advance in our understanding of nature. Zeno's paradoxes do not really fall into this category, because his logic as well as his assumptions were faulty, but Zeno's paradoxes are well worth discussing.

Most of us take motion for granted, but not everybody. Some of the ancient Greek philosophers thought that motion was as illusion, and tried to prove it. Foremost among these philosophers was Zeno of Elea (c. 490–c. 430 BCE), a follower of the Elean philosopher Parmenides. Parmenides had argued that all truth was *in being*, and that change was therefore impossible. Zeno, true to his school of philosophy, tried to show logicially that motion was an illusion which could not exist in reality.

In order to prove that motion was impossible, Zeno posed some problems concerning motion, which apparently led to contradictions. These problems are known as Zeno's paradoxes. We discuss two of them.

Achilles and the tortoise. Achilles, who is very fleet of foot, and a tortoise, who is very slow, have a race. Let us suppose that Achilles can run ten times as fast as the tortoise, but the tortoise has a 100 meter head start. Can Achilles catch the tortoise? Our common sense tells us that of course Achilles can catch the tortoise, but Zeno argues to the contrary as follows: In the time it takes Achilles to travel the 100 meters to reach the original position of the tortoise, the tortoise will have gone 10 meters. While Achilles travels the 10 meters separating them, the tortoise will travel 1 meter. While Achilles travels the 1 meter the tortoise will travel 0.1 meters (one-tenth meter) and so on. Whenever Achilles reaches the point where the tortoise *was*, the tortoise is no longer there, but is farther on. By this same reasoning, nobody can catch anybody in a race.

The arrow. A man shoots an arrow from a bow. Will it move? Again our common sense tells us that of course the arrow will move, although it might not hit its intended target. But Zeno argues as follows that the arrow will not move: Before the arrow can travel the whole distance to its target, it must go half the distance. Before it can go half the distance, it must go a quarter, before a quarter, an eighth, and so on. Carrying this reasoning on and on, Zeno argues that the arrow cannot get started at all. And what is true for an arrow, is true for anything else, so all motion is an illusion.

We all know that Zeno's arguments lead to conclusions that are contrary to observation, and therefore must be wrong. But do we know why his reasoning is wrong? Zeno's contemporaries did not know. It was after Newton and Gottfried Wilhelm von Leibniz (1646–1716) independently invented the branch of mathematics known as calculus in the 17th century that an apparently satisfactory refutation of Zeno was made.

The calculus can handle problems, such as Zeno's, which involve a number of operations which grows without end, or, in other words, becomes infinite. Zeno breaks down the race between Achilles and the tortoise into an infinite number of parts, but it does not follow that the race will take an infinite amount of time. This is because as the distance involved in each successive part becomes shorter, the time to carry it out also decreases proportionately. The calculus enables us to show that an infinite number of distances can be traversed in a finite amount of time, if the distances traversed are sufficiently small, as is the case with Achilles and the tortoise. Achilles has to travel only a finite distance to catch the tortoise, and so it will take him only a finite amount of time to do it, even if somebody gives a prescription for subdividing the distance into an infinite number of parts. The resolution of the paradox of the arrow follows similar reasoning.

We really do not need the full power of the calculus to resolve Zeno's paradoxes. Zeno breaks up the distance traveled by Achilles into a sum of an infinite number of terms, which form what is called a "geometric series." There is a simple formula which gives the finite sum of this series (we do not write it down here). So Achilles has to travel only a finite distance to catch the tortoise, and with a sufficiently large speed he will be able to do it in a reasonable amount of time.

The calculus has apparently enabled people to show that our common sense ideas about motion do not involve any paradoxes. But wait! Perhaps we have not yet heard the last word on the subject. Matter was once thought by some people to be infinitely divisible, so that, in principle, a piece of gold, could be divided again and again without end, and still the bits would remain gold. Now, however, we know that gold and other substances are made of atoms, which cannot be subdivided without altering their properties. In other words, if we can figure out a way to continually subdivide a piece of gold, eventually it will cease to be gold. But what has this to do with Zeno's paradoxes? Just this: perhaps space itself cannot be infinitely subdivided, although we currently have no evidence in this regard. If it should turn out that there is a limitation on the divisibility of space, then Zeno's hypothesis that he can subdivide the distance without end would be in error. Zeno's paradoxes would again be resolved, but the resolution would involve ideas which are quite different from those most of us believe today.

4.2 Aristotle's views on motion

Aristotle not only had a picture of the universe but also had views on motion. Again, it was Galileo who showed that Aristotle's ideas were basically wrong. Although not many of Aristotle's views on science have stood the test of time, it is in the nature of science that old views are often superseded by new discoveries.

Much of Aristotle's importance in physics derives from his emphasis on direct observation of nature and his teaching that theory must follow fact. This was a profound departure from the view of many earlier Greek philosophers (and many later philosophers from various parts of the world) that understanding could be obtained from pure reason.

Today, many scientists believe in a synthesis of the views of Aristotle and his opponents. They believe that, although observation is a necessary ingredient in science, observation does not by itself lead to the discovery of nature's laws. According to this view, flights of imagination are also essential. Just as knowledge of the grammar of a language does not by itself make a great writer, knowledge of the "scientific method," whatever it may mean, does not make a great scientist. There is an art to being a great scientist.

Now let us turn to Aristotle's physics. He believed that on earth the natural state of an object was to be at rest. We can easily understand how he got that idea. Roll a ball along the floor and it comes to rest. Slide on the ice and you come to rest. Then how did Aristotle explain the fact that motion exists on earth? According to Aristotle, motion on earth is a transitory phenomenon. Objects can be set in motion, but, left to their own devices, sooner or later they will come to rest.

Aristotle also taught that the natural state of motion of objects in the air above the earth was to fall at constant speed. Raindrops obey Aristotle's law to a good approximation. Aristotle further taught that heavy objects fall faster than light ones. For example, a stone falls faster than a feather.

Aristotle must have noticed that some falling objects do not fall at constant speed but accelerate. We have already noted that acceleration is defined as the rate of change in velocity. Another way of putting this is to say that the acceleration is the change in velocity divided by the change in time. Hold an object stationary in your hand, and then let it go. It must accelerate for a time, or it would remain at rest suspended in the air, contrary to the observation that it falls to the ground. Aristotle must have regarded the initial acceleration of a falling object as transitory, as he regarded motion on earth as transitory. Sooner or later, Aristotle must have thought, the object would stop accelerating and fall at constant speed, provided it did not hit the ground first.

Although Aristotle was right in many of his observations, he drew the wrong conclusions from them, or rather, more precisely, he failed to understand the essential properties of motion. A simple example will show what Aristotle missed. Take a run and slide on a gravel path. Take a similar run and slide on a patch of smooth ice. It is true that in either case you come to rest. But you slide much further on the ice than on the gravel. Why? Aristotle evidently did not think the answer important. But if we want to go beyond merely noting that objects come to rest on earth, we must face the question of why under different circumstances objects take different times to come to rest.

We here briefly anticipate the answer that Galileo gave, which we shall discuss in more detail in the next section: A person sliding on gravel stops sooner than he does when sliding on ice because the gravel gives more *resistance* to the motion than does the ice. Once we have the idea that objects come to rest because of resistance to their motion, we might believe that, contrary to what Aristotle thought, motion tends to persist unless opposed.

Consider another example. Drop a balloon and a stone from the top of a building. The balloon falls slowly, most of the time at constant speed, but the stone accelerates all the way to the ground. Why? Again, Aristotle did not have a satisfactory answer. And again Galileo realized that it was resistance, in this case the resistance of the air, that causes the motion of the ballon and the stone to be so different. We shall discuss Galileo's answer further in the next section.

4.3 Galileo on motion

Galileo is supposed to have dropped a large stone and a small stone from the leaning tower of Pisa and noticed that, within experimental error, the two stones hit the ground at the same time. Whether or not Galileo actually performed this experiment, he had profound and original ideas about motion, some of which we still believe to be true.

According to Galileo, objects fall, not with constant velocity, but with constant acceleration. This acceleration is the same constant for both light and heavy objects. Therefore, if two stones, one light and one heavy, are dropped from the same height at the same time, they will strike the ground at the same time. But this idea cannot be correct for all objects, because, as we well know, if a stone and a feather are dropped from the leaning tower of Pisa (or any other building), the stone will hit the ground first.

Galileo was well aware of this objection, for it is nothing more than the objection of Aristotle. Galileo realized that it is only the resistance of the air, which is more important for the feather than for the stone, that prevents a feather and a stone from falling with the same acceleration. The air can have even more drastic effects, lifting objects from the ground and enabling birds to fly. If we are not careful, the existence of the air will divert us from understanding the essential motion of falling bodies. Take away the air, and the stone and feather will strike the ground simultaneously, and birds would not be able to fly.

A variation of this experiment is frequently done in elementary physics courses. A feather and a stone are put in a long, glass tube, closed at one end and containing a valve at the other end. When the tube is held vertically, the stone falls from top to bottom much faster than the feather. The tube is then attached to a vacuum pump (a device for removing gases, such as air, from a container) and nearly all the air is removed. The valve is then closed, and again the tube is held vertically. Both the stone and the feather then fall with the same acceleration and strike the bottom of the tube at the same time (applause from the audience). This demonstration is shown in Figure 4.1.

Galileo could not take away the air from around the leaning tower of Pisa, but was able to perform experiments with objects like large and small stones, for which the effects of air resistance are small. He then was able to reason that, in the absence of air, *all* objects would fall with the same acceleration. Time has borne out the correctness of

Galileo's reasoning to within the very small errors of modern experiments.

Figure 4.1: Illustration that a stone and feather fall with the same speed and the same acceleration in a closed glass tube from which almost all the air is removed.

Galileo also had profound ideas about horizontal motion. Unlike Aristotle, Galileo realized that the "natural" tendency of an object in motion is to move at constant velocity (i.e., at constant speed in a straight line), and it is only external influences (forces) which change the speed, the direction of motion, or both i.e., cause acceleration. The *inertia* of a body is its tendency to resist acceleration. Galileo enunciated the *law of inertia*, which states:

A body at rest will remain at rest and a body in motion will remain in motion at constant velocity unless disturbed by an external influence.

No matter how fast a person is moving horizontally, if she is moving at constant velocity, she will not feel any effect of the motion (provided that the air moves with her). When we are at rest with respect to the earth, we are actually moving very fast along with the earth. Even though the earth is rotating, it takes 24 hours to rotate once. Furthermore, the earth is very large on the human scale. As a consequence, when we move along with the earth, we are moving approximately in a straight line. The curved motion is too slight for us to feel, and that is why it is hard to tell that the earth is moving and that we are moving along with it.

Let us take an example of acceleration. If you are driving a car and want to stop for a red light, you know that usually it is not enough to remove your foot from the accelerator pedal; you must put your foot on the brake. The influence of the brake causes the car to decelerate faster than the influence of air and other ordinary resistance (friction) to the motion of the car.

The above example of a car's brake may cause some confusion. How can the brake of a car be regarded as an external influence? Actually, a brake is an internal influence, causing the wheels of the car to rotate more slowly. Whether this effect influences the car to slow down depends on external influences. Under ordinary circumstances, the friction of the road on the car's wheels causes the car to slow down when the brake is applied. However, if the car is traveling on an icy road, friction is reduced and the wheels will slide even if they do not rotate. An extreme example is if the car is driven off the side of a bridge. Applying the brakes has no effect on the overall path of the car while it is in the air.

Gravity is the influence which causes objects to accelerate downward when dropped. But gravity is a remarkable influence in that, in the absence of other influences, it gives all objects the *same* acceleration. Galileo knew this, but did not give any explanation for it. We shall see in Chapter 6 that Newton was able to account for it in terms of his universal law of gravitation and his laws of motion, but he did not discuss a possible underlying meaning behind the law of gravity. As we shall see in Chapter 10, Einstein thought deeply about the subject and came up with a theory that was different from Newton's theory of gravity. Subsequent experiments showed that Einstein's theory was closer to the truth.

Chapter 5

Newton's Laws of Motion

> I do not know what I may appear to the world; but to myself I seem to have been only like a boy playing on the seashore, and diverting myself in now and then finding a smoother pebble or a prettier shell than ordinary, whilst the great ocean of truth lay all undiscovered before me.
>
> —Isaac Newton (1642–1727)

5.1 The first law

The law now known as Newton's first law of motion is really Galileo's law of inertia. It is so fundamental as to bear repeating:

A body at rest will remain at rest and a body in motion will remain in motion at constant velocity unless acted upon by an external force.

This statement of the first law uses somewhat different language from that used in the last chapter. What is called "an external influence" in the previous chapter is here called "an external force." We have defined a force as a push or pull. The essential feature of the first law is that motion tends to persist uniformly, i.e., at constant velocity, which may be zero velocity. It takes a force to change either the magnitude or direction of the velocity of a moving body. When a force acts on a body, it acts in a certain direction. Because force has both magnitude and direction, it is a vector. The branch of physics that deals with analyzing the action of forces on matter is called "mechanics."

Not only does it take a force to change the velocity of a body, but the body resists the change. The amount of resistance of the body is called its *inertia*. You do not have to push very hard on a rolling ball to give it a big deflection from its path. If you exert the same amount of force on a moving car (not recommended!) you will change its path very little—you will not even notice the change in the car's motion. The reason for the difference is that the car has much more inertia than the ball. The inertia of a body is evidently related to the amount of matter in the body.

A concept closely related to force is "pressure," which is defined as the amount of force acting on a certain area of a body divided by the area, or, in other words, pressure is force per unit area.

5.2 The second law

The law of inertia says that a force is necessary to change the velocity of a body, that is, to accelerate a body. Newton went beyond this qualitative idea to make it quantitative. Newtons's second law states:

The acceleration of a body is *proportional* to the external force acting on it and *inversely proportional* to its *mass*.

The mass of a body is the "quantity of matter" it contains. The mass is a scalar, and we restrict ourselves here to the case in which the mass is greater than zero. In a later chapter we discuss the case of an object with zero mass. As far as we know, the mass of a body cannot be negative. Because the inertia of a body is its resistance to being accelerated, the mass of a body is a quantitative measure of its inertia. The larger the mass of a body, the larger is its inertia and the smaller its acceleration when subject to a given force.

While we are on the subject of mass, we introduce the related concept of density, which is defined as mass per unit volume. If we have two pieces of iron, one of which has twice the volume of the other, the bigger piece has twice the mass of the smaller piece, but both have the same density. On the other hand, a piece of iron has a larger density than a piece of stone, independent of the sizes of the two pieces. Wood floats on water because its density is lower than that of water, while a stone sinks because its density is higher than that of water.

We now make more precise what proportionality means. If the acceleration is proportional to the force, it means the acceleration is equal to a constant times the force. If the acceleration is inversely proportional to the mass, it means the acceleration is equal to a constant divided by the mass. If we double the force acting on a body, we double its acceleration; if we multiply the force by 3, the acceleration gets multiplied by 3, and so on. Also, if we apply the same force to two bodies, one of which has twice the mass as the other, the heavier body will undergo only 1/2 as much acceleration as the lighter body.

We can combine the two ideas of Newton's second law by writing it in symbols, letting the symbol **F** denote the force, the symbol m denote the mass, and the symbol **a** denote the acceleration. Newton's second law then says that the acceleration equals the applied force divided by the mass, or $\mathbf{a} = \mathbf{F}/m$, or equivalently $\mathbf{F} = m\mathbf{a}$. In these equations, the units of force, mass, and acceleration are chosen so that the constant of proportionality is unity. We show in in Figure 5.1 that the acceleration is in the same direction as the force.

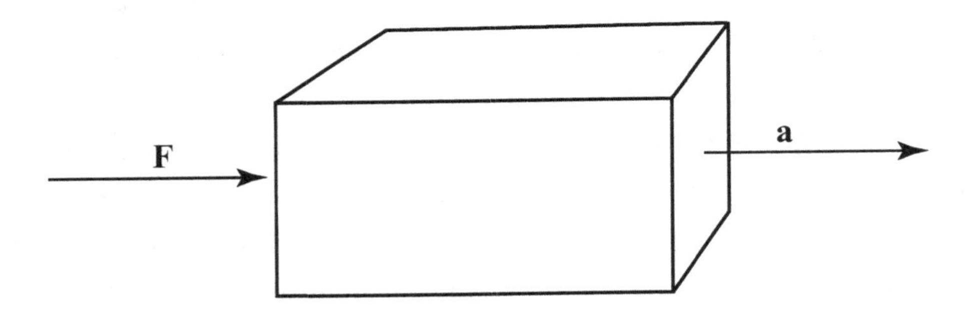

Figure 5.1: Illustration of the fact that the force **F** and the acceleration **a** are in the same direction.

Because both sides of an equation are equal, the units of the left side should equal the units of the right side. Therefore, the unit for force is equal to the unit for mass multiplied by the unit for acceleration. In the metric system, the primary unit of mass is the kilogram (kg), about 2.2 pounds. The unit for acceleration in the metric system is meters/second/second (m/s^2). Therefore, the unit of force is kg m/s^2. One kg m/s^2 is called a newton, and abbreviated N. Often in the metric system we use the gram, abbreviated g, for mass instead of the kilogram. It takes 1000 g to make 1 kg. We also use the centimeter, abbreviated cm, instead of the meter. It takes 100 cm to make 1 m.

Often Newton's second law is stated in words in the following form:

> The force on a body is equal to its mass times its acceleration.

It should be understood that **F** is the *net external* force acting on the body. The net external force is the vector sum of all the forces acting on the body from outside the body. If two forces, equal in magnitude but opposite in direction, act on the same part of a body, their effects cancel each other and the net force is zero. Therefore, the body will not accelerate. However, two forces acting on different parts of a body can accelerate the parts relative to one another. Such forces can cause a body to rotate at increasing speed, for example. A solid body has internal forces which tend to hold it together, but if the external forces are too strong, the body can be crushed or torn apart.

In the previous chapter be have discussed a person sliding on gravel and coming to rest because of an influence opposing the motion. Now we can identify the influence as a force, called a frictional force. Friction acts because of the roughness of the two surfaces that slide across each other. Surfaces cannot be made completely smooth because of the atomic nature of matter, which we shall discuss in Chapter 11.

A note of caution is in order about Newton's second law. This law, in the form we have stated it, is only approximately true. The twentieth century theories of relativity and of quantum mechanics require that the law be modified in certain circumstances. In the case in which bodies move very fast, Newton's second law is superseded by Einstein's theory of special relativity. In the case in which forces act on microscopically small objects, such as atoms, the second law is superseded by the laws of quantum mechanics.

Physicists often call forces "interactions." The idea of an interaction is more general than that of a force because an interaction not only can cause an object to accelerate but can create or destroy (annihilate) particles. Creation and annihilation take place at the microscopic level, and, as far as we know, the laws of quantum mechanics apply.

We shall have more to say about relativity and quantum mechanics in later chapters. In most ordinary circumstances (for macroscopic objects not moving too fast), Newton's second law is an excellent approximation to the way nature behaves, so good in fact, that it is extremely difficult to measure deviations from it.

Newton's first law (the law of inertia) is just a special case of his second law. If the force is equal to zero then the acceleration must also be equal to zero. But if the acceleration is zero, then the velocity is not changing. Therefore, the velocity of the body must be constant, as the law of inertia requires in the absence of a net force on the body. Note that if the body is at rest, the velocity is zero, which is a particular constant.

Let us now give a few applications of Newton's second law. If you push a stone along a level sidewalk and then let go, the stone soon comes to rest. We can understand the motion of the stone in terms of the forces acting on it. Originally, the stone is at rest in your hand. In order to get the stone moving, you supply a force to it with your hand. Once you let go of the stone, you no longer supply any force to it. If no other forces acted on the stone, it would continue to move in a straight line at constant velocity. However, it slows down (accelerates in the direction opposite to the direction of its velocity) and stops. Therefore, another force must act on the stone while it is slowing down. This force is the frictional force which the sidewalk imparts to the stone. The smoother the sidewalk, the less the frictional force, and the farther the stone will travel before coming to rest.

As a second example, if you drop a ball from rest, it accelerates on the way down. (If it did not accelerate, it would remain motionless in mid air after you let go of it.) According to Newton's second law, a force is necessary to give the ball acceleration. In this case, the force is the force of gravity. We shall have much more to say about gravity later.

5.3 The third law

Newton's third law of motion states:

Whenever one body exerts a force on a second, the second body exerts an equal and opposite force on the first.

In this statement, the word "equal" means equal in magnitude, and the word "opposite" means opposite in direction. Note that the two forces mentioned in the third law act on different bodies.

The third law is a statement about the relationship between two forces at a given instant in time. If the force the first body exerts on the second changes as time goes on, so does the the force the second body exerts on the first.

As an example of the third law, suppose you hold a suitcase in your hand. You are exerting an upward force on the suitcase to balance the force of gravity acting downward on it. But because your hand is exerting an upward force on the suitcase, the suitcase is exerting a downward force of the same magnitude on your hand. Your hand does not accelerate downward because of a balancing upward force your muscles exert on your hand to keep it at a constant height. If the suitcase is heavy, your muscles will get tired from the upward force they exert.

As another example, suppose you shoot a gun. Because the gun exerts a forward force on the bullet, the bullet must exert a backward force of equal magnitude on the gun. This backward force causes the gun to recoil.

As a third example, the engines of a jet airplane exert a backward force on hot gases which arise from burning the fuel. Therefore, the gasses exert a forward force on the airplane, pushing it forward. In steady flight, the airplane does not accelerate forward because the resistance of the air is a backward force on the airplane which cancels the forward force exerted by the hot gases. in Figure 5.2 we illustrate that the force $\mathbf{F_g}$ on the gases is equal in magnitude and opposite in direction to the force $\mathbf{F_a}$ of the gases on the airplane.

Newton's third law holds for contact forces, by which we mean

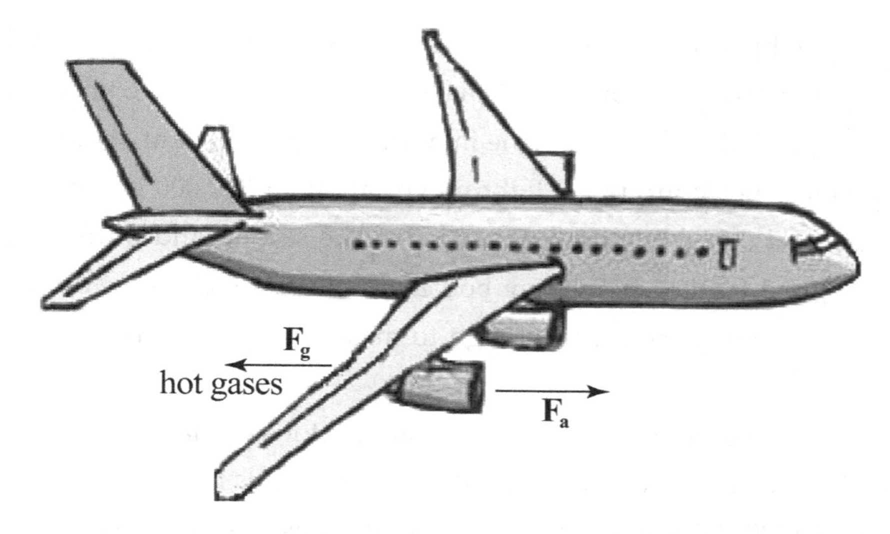

Figure 5.2: Illustrating that the force $\mathbf{F}_g$ of a jet airplane on hot gases is equal in magnitude and opposite in direction to the force $\mathbf{F}_a$ of the gases on the airplane.

forces between two bodies which are touching each other. However, the third law is only approximate for forces, like gravitational and electrical forces, which act between two separated bodies. If the bodies are moving rapidly with respect to each other, an improved description is given by Einstein's theory of relativity, which we discuss in Chapter 10.

5.4 Frames of reference

As far as we know, Newtons's first law (the law of inertia) is exact. However, we know that Newton's second and third laws are only approximate, although they hold very well for slowly-moving macroscopic bodies. By slow, we mean slow compared to the speed of light, which is very fast indeed. An airplane moves fast compared to a person walking, but the airplane moves very slowly compared to the speed of light. A jet plane in normal flight moves at about 900 kilometers per hour, while light moves at about 300,000 kilometers per second (186,000 miles per second). The second and third laws are so good for slowly-moving bodies that in many cases we do not have instruments sensitve enough to measure any deviations from them.

We digress here to introduce a notation for very large and very small numbers, as we encounter them often in this book. The speed of light (300,000 km/s) can also be written as 3×10^5 km/s. The notation 10^5 means 1 followed by 5 zeros, or 100,000. Similarly, 10^8 means 1 followed by 8 zeros or 100,000,000 (one hundred million). We use negative powers of 10 for very small numbers. For example, the number 10^{-4} means 1 divided by 10,000, or 0.0001,

In discussing Newton's laws, we have to point out a certain limitation in their validity that has nothing to do with speed. Let us concentrate our attention on the law of inertia, which is exact independent of speed. Nevertheless, there is an important limitation to the law of inertia in that it is observed to hold only when measurements are made from a so-called *inertial frame of reference.*

What is a frame of reference, and what is an inertial frame? Before defining these notions, we illustrate them with an example. Go outside on a clear night and look up at the stars. Even though they may be moving rapidly, they seem motionless because they are very far away. That is why they are often called the "fixed" stars. However, suppose you rotate your body while looking at the stars. Then, if you consider yourself at rest, the stars make circles overhead–they are no longer motionless from your point of view. If I remain at rest and watch both you and the stars, then, from my point of view the stars are fixed (not moving) and you are rotating. Who is to say whose point of view is "better," yours or mine?

A "point of view," used in this sense, is called a "frame of reference." Sometimes, we shorten this simply to "frame." According to modern ideas, both your frame of reference and mine are equally good; this is a principle of general relativity, proposed by Einstein and discussed in Chapter 10.

It turns out that Newton's first law (the law of inertia) holds for a person making a measurement if that person is in the frame of reference in which the stars are fixed (motionless). For that reason, such a frame of reference is called an "inertial frame." Furthermore, it turns out that any frame of reference which is moving at constant velocity with respect to the fixed stars is also an inertial frame. However, a frame of reference that is accelerating with respect to the frame of the stars is not an inertial frame. If you turn in a circle, you are changing your direction, which means you are accelerating, and so you are not in an inertial frame.

Even to a person who is stationary on earth, the stars do not appear to be absolutely at rest. This is because the earth rotates on its axis once a day, so the stars make apparent circles in the sky once a day. Because the motion is not in a straight line, it is accelerated motion. The apparent rotation of the stars shows that the frame of the earth is not strictly an inertial frame. This implies that an observer who is stationary on earth is not strictly in an inertial frame. However, because the earth is large and rotates on its axis only once a day, the earth is very nearly in an inertial frame. In fact, for many measurements, we can neglect the fact that the earth is rotating and use Newton's laws of motion without modification.

There are some measurements, however, that clearly show that the earth's frame is rotating. For example, consider a pendulum suspended from a high ceiling by a wire. If you give the pendulum a push, it will swing back and forth in a straight line. Make a chalk line on the floor parallel to the direction the pendulum is swinging. If you leave the room for a few hours and then return, you will find that the pendulum no longer swings parallel to the chalk line, but makes an angle with respect to it. The force that causes the direction of the swing to change is a "fictitious" or "inertial" force. From the point of view of an observer in an inertial frame (not rotating with the earth), the pendulum does not change its direction of swing at all, but the earth rotates underneath it.

Fortunately, physicists are able to take account of the earth's rotation and make more precise calculations than would be possible if the earth's frame is taken as an inertial frame. Nevertheless, unless we state otherwise, we shall regard the earth's frame of reference as an inertial frame. In doing so, we are making an approximation, and our results will have an error. Physicists frequently make approximations to simplify calculations. A physicist usually has a good idea of the size and importance of the errors he makes as a result of his approximations. If the errors are unimportant, the approximations are useful.

Physicists occasionally make approximations even if they know the resulting errors will be big, or if they don't know how important the errors will be. Physicists may make such approximations either because they will be satisfied with a rough estimate of the value of a quantity or because they cannot solve the problem exactly. The philosophy in the latter case is that a poor answer is better than no answer at all. (This is not always true.)

There are also errors in measurements, so even when a physicist measures a quantity, he obtains only an approximate answer. When we say a calculation (prediction) from a theory and the result of a measurement "agree," we mean only within the errors of the measurement and the calculation.

The "rest frame" of an object is the frame in which it is at rest. An observer makes observations and measurements from his own rest frame, no matter how he may be moving with respect to another frame, such as the frame in which the earth is stationary or the frame in which the stars are fixed. The rest frame of an object may or may not be an inertial frame, depending on how the object is moving when viewed from an inertial frame.

It has been observed that the frame in which the stars are fixed is an inertial frame and that all frames of reference which move with constant velocity with respect to the frame of the fixed stars are also inertial frames. Because of this, we cannot use the law of inertia to determine whether we are at absolutely at rest or whether we are moving at constant velocity. From the point of the law of inertia, absolute motion at constant velocity is not observable, and we can measure only relative motion. However, there is a way to measure the local rest frame of the universe. We cannot measure a global rest frame of the universe because the universe is expanding, as we shall discuss in Chapter 19.

If you ride in a car at constant velocity, you do not feel any forces on you. Therefore, if you close your eyes, you cannot tell how fast you are moving or even whether you are moving at all. (Actually, you can tell that you are moving because no car can travel with strictly constant velocity. Small forces cause the car to vibrate when it is moving, and vibration is accelerated motion.) If you open your eyes and look out the window, you see the trees at the side of the road rushing backwards. Because you know that the trees are stationary on the earth, you know you are moving with respect to the earth, or, alternatively, that the earth is moving with respect to you.

We next consider accelerating frames of reference. By an accelerating frame, we mean a frame which is acclerating with respect to the fixed stars. A rotating frame is a special case of an accelerating frame.

It is easy to see that the law of inertia does not hold in accelerating frames. Slide a smooth stone over a smooth floor, so that friction can be neglected (to a good approximation). In this approximation, there are no forces acting on the stone and so it slides with constant velocity, in accordance with the law of inertia. However, while you are watching the stone, accelerate your own body. If you are accelerating with respect to the stone, the stone is accelerating with respect to you in the opposite direction. Therefore, When measured from your frame, the stone accelerates even though no forces are acting on it, in violation of the law of inertia. Thus, the law of inertia is violated when measured from an accelerating (noninertial) frame.

It is apparant from this example that not only does Newton's first law fail to hold in an accelerating frame, but also his second law. After all, the first law is a special case of the second, so that if the first law fails, so does the second. The third law, however, holds to a good approximation in accelerating frames, especially for contact forces.

Newton's first two laws can be modified to hold in an accelerating frame only by introducing so-called "fictitious" forces to account for the acceleration of objects measured with respect to the accelerating frame.

As an example of a "fictitious" force arising in an accelerating frame, consider what happens when you turn to the left in a car while traveling at high speed. (This is an acceleration by virtue of a change in direction.) The answer is that you will feel yourself pushed to the right, and if you don't wear your seat belt, you may actually slide along the seat to the right. Although the force that pushes on you to the right seems real enough to you, it is regarded as a "fictitious" force from the point of view of an observer in an inertial frame. According to the observer in the inertial frame, the car turns to the left because of a force on it in the *left* direction arising from the friction of the road on the wheels. According to the same observer, you are also pushed to the *left* by the seat belt which is attached to the car. The observer cannot identify any force on you to the *right*. If you say you feel a force to the right, he will say it is a fictitious force because you are not in an inertial frame. It is your inertia that resists the force pushing you to the left and feels like a force pushing you to the right. Therefore, sometimes the "fictitious" forces are called "inertial" forces. Neither word for the extra force felt by someone in a non-inertial frame is a good one. The term "fictitious" is not good for a force that feels as real as any other force. The term "inertial" is not good because the force is absent when measured from an inertial frame of reference. But "fictitious" and "inertial" are the words commonly used, and so we use them in this book.

The conclusion of all this is that observers making measurements from accelerating frames need to include forces which seem real enough from their own point of view but are absent from the point of view of an observer making measurements from an inertial frame.

Chapter 6

Newton's Theory of Gravity

> If I have seen farther than others, it is because I have stood on the shoulders of giants.
>
> —Isaac Newton

6.1 An apple and the moon

Newton showed that the motion of an apple falling to earth and the motion of the moon around the earth arise from the same force of gravity. In order to understand how the motion of an apple in a straight line and the motion of the moon in an approximate circle (actually an ellipse) can arise from the same force law, we first examine motion on earth in some detail.

Gravity is the force that makes objects fall to earth. But these objects do not always fall in a straight line. It is true that an object dropped from rest, like the apple, will fall straight down. However, if you throw a ball horizontally, its subsequent motion in the air will be a combination of the horizontal motion you originally gave it and vertical motion due to gravity. The actual path is curved, and is approximately given by the curve called a parabola. in Figure 6.1 we illustrate the path of a ball thrown initially with horizontal velocity.

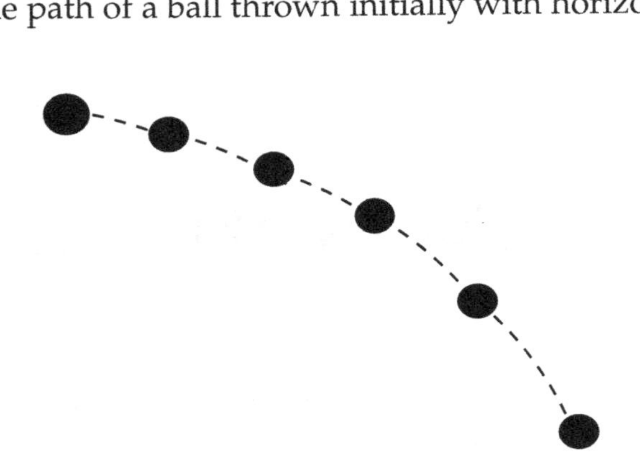

Figure 6.1: Parabolic path of a ball thrown with initial horizontal velocity. Pictures of the ball are given at equal time intervals.

The horizontal component of the ball's velocity will decrease slightly because of air resistance, while the vertical component of the velocity will increase because of the downward force of gravity. The net result is that the ball starts out horizontally, but falls at a steeper and steeper angle from the horizontal until it hits the ground.

Imagine throwing the ball horizontally in the absence of air resistance. The horizontal component of the velocity will remain constant while the vertical component increases because of gravity until the ball hits the ground. Then imagine throwing the ball at a greater initial speed. It will of course travel farther before it hits the ground. One can conceive of throwing the ball at such a high speed that it never hits the ground (neglecting irregularities in the earth such as mountains). The ball curves toward the earth, but the earth is also curved so that a sufficiently fast-moving ball can remain a constant distance above the surface of the earth even as it continues to fall. This is approximately the situation with the moon, except that it orbits the earth in an ellipse rather than in a circle. The moon keeps falling toward the earth, but the horizontal component of its velocity is sufficiently large that it it remains in orbit above the surface of the earth.

Newton guessed that the force of gravity between two pointlike objects is proportional to the product of their masses and inversely proportional to the square of the distance between them. It is an attractive force. The constant of proportionality is called the "gravitational constant" and is denoted by the symbol G. In symbols, the magnitude of the force is given by $F = Gm_1m_2/r^2$, where m_1 and m_2 are the masses of the two objects and r is the distance between them.

Newton was not the only one to guess the inverse square law, but he showed that the law was universal, acting on the moon and the apple with the same coupling constant G. In order to show this, he had to assume that the earth attracts the apple as if all the earth's mass were concentrated at its center. Later he proved this assumption by taking the earth to be a sphere and summing up the gravitational contributions from every bit of matter in the sphere.

Newton did not know the mass of the earth or moon, and he did not know the value of the gravitational constant G. However, he assumed the same value of G and the same velue of the mass of the earth in both cases, so that by taking the ratio of the forces (dividing one by the other) the quantities G and the mass of the earth cancel out and so do not need to be known.

Newton also did not need to know the mass either of the moon or of the apple because of his second law of motion, which says that the acceleration of a body is inversely proportional to its mass. Because the acceleration is obtained by dividing the force by the mass and because the gravitational force is proportional to the mass, the body's mass cancels out in the expression for the acceleration.

The net result of all of this is that the ratio of the acceleration of the moon to the acceleration of the apple equals the inverse ratio of the squares of their distances to the center of the earth. In order to calculate the ratio, Newton had to know the radius of the earth and the distance to the moon, numbers that he was well aware of. He then compared his result with measurements of the accelerations of the moon and the apple, and found that they agreed with his calculation.

The fact that acceleration of a body due to gravity is independent of its mass is why Galileo found that a large stone and a small stone fall with the same acceleration.

6.2 Action at a distance

Newton invented the calculus in part because it proved a useful tool for him to calculate the motion of the planets using his formula for the force of gravity together with his second law of motion. The results were spectacular: Newton was able to show that Kepler's three laws of motion were consequences of the universal law of gravitation.

Despite the great success and fame that Newton achieved, he was troubled by an aspect of gravity that he could not understand: action at a distance. We can appreciate Newton's problem if we compare the gravitational force with another kind of force, for example, a push. If someone pushes you, the person's hand makes contact with your body, and you directly feel the force on your body. However, the force of gravity appears to act through empty space across the vast stretches of the solar system and beyond. Furthermore, you don't feel the force of gravity in the same way that you feel an ordinary push. If you jump off the roof of a building, while you are in free fall, you feel no force on you (except for air resistance). It's not the fall that hurts you; it's the sudden stop when you hit the ground. If you are in a rocket ship orbiting the earth (in free fall), not only do you not feel your motion, but you are weightless. You can float inside the ship without feeling any force on you. Clearly, there is something different about gravity, but Newton didn't know what it was. In Chapter 10, we shall see how Einstein modified Newton's theory of gravity in such a way as to explain its previously mysterious properties.

6.3 Fields

A partial solution to the problem of action at a distance was given by Michael Faraday (1791–1867), an English physicist, who introduced the concept of a field in connection with electricity and magnetism (see Chapter 8). Let us first look at a field in a simple case, that of temperature. The temperature at one place is a scalar: only one number is required to specify it. But to specify the temperature more generally, one needs to specify it at every point in space, because it is not the same everywhere. Likewise, the temperature changes with time. A quantity that has a value at every point in space and time is called a "field." The temperature field is a scalar field because only one number specifies it at each point in space and time.

However, there can also be a vector field, such as a force field, which has direction as well as magnitude. To specify a force field completely, three numbers must be given at each point in space and time: one number to give its magnitude and two more numbers to give its direction. Let us introduce a gravitational field. In Newton's theory of gravity, one can imagine a vector gravitational field existing at each point in space such that the gravitational force on a test mass at that point is equal to the product of the mass times the field. The gravitational field at every point results from the combined effect of all the masses other than the test mass. If there is only one other mass, the gravitational field obeys in inverse square law.

The idea of a gravitational field does not really *explain* action at a distance. However, if we think of space itself as not being really empty but containing a real gravitational field, we might be somewhat less astonished that a mass anywhere in the field will feel a force on it. In Einstein's theory of gravity (see Chapter 10), the gravitational field is not a vector field but is a more complicated quantity called a "tensor" field. More than three numbers at every point in space and time are required to specify a tensor field like the gravitational field. In common situations, where Newtonian gravity is an excellent approximation, the extra components of the gravitational field are negligible.

Chapter 7

Energy and Momentum

> The fundamental concept in social science is Power, in the same sense in which Energy is the fundamental concept in physics.
>
> —Bertrand Russell (1872–1970)

7.1 Work

"John works hard learning physics." In the preceding sentence, the word "work" is used in a popular sense. Physicists, on the other hand, use the word "work" with a specialized meaning which is quite different. In physics, work is done by a force which causes an object to move.

If an object moves in a certain direction because of an applied force, then the work done on the object by the force is defined as the component of force in the direction of motion multiplied by the distance the object moves.

Let the distance be s and the component of force in the s direction be F_s. Then, in symbols, the work W is given by $W = F_s s$. If the force component F_s is not constant during the time the particle moves, then the distance s must be broken into segments which are small enough that the force does not change appreciably while the object moves along the segment. The partial work done during the motion along a segment is found by multiplying the segment by the value of the force component while the object traverses that segment. The total work is found by adding up the work done during each segment. Because this is a complicated process involving calculus, we concentrate on constant forces. In Figure 7.1 we show the work done by a constant horizontal force of magnitude F in moving a block of wood a distance s. The work W done is $W = Fs$.

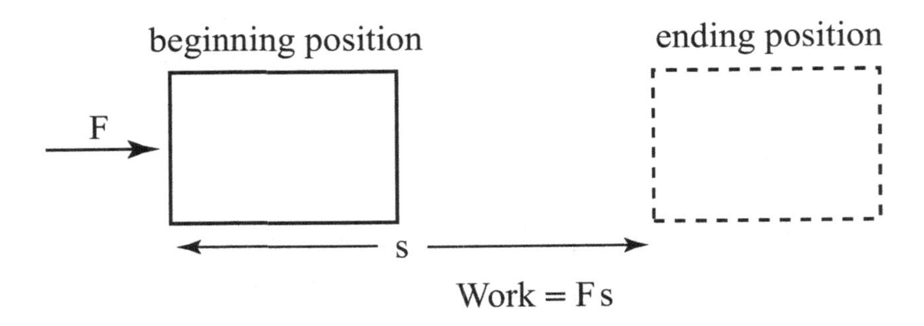

Figure 7.1: Work done by a horizontal force of magnitude F pushing a block of wood a distance s.

Suppose you hold a ball stationary in your hand. You are exerting an upward force on the ball to counterbalance the downward force of gravity on it. However, you are not doing any work because the ball does not move. Now, suppose you walk across a room with the ball, taking care that the ball does not move up or down but only horizontally. You are still doing no work on the ball, because the force has no component in the direction the ball moves. But if you lift the ball, you are doing work because the ball moves in the same direction as the force you exert.

If you push a table across the floor, you are doing work because the force is in the same direction that the table moves. Suppose you push the table at constant velocity. There must not be any *net* force on the table because it does not accelerate. The reason the net force is zero is that an opposing frictional force cancels the force that you exert on the table. Nevertheless, you are doing work on the table.

7.2 Energy

In the previous section we discussed an example in which you do work by pushing a table across the floor. You are able to do this because you posess within you an attribute that physicists call "energy." If you had no energy, you would not be able to do any work. So in some sense, *energy is the ability to do work*. This is not the whole story, however, because there are limitations on the ability of energy to do work.

Energy comes in many different forms, to which physicists give various names. Among the many forms of energy, we mention a few now (and define them later in this chapter): kinetic energy, potential energy, heat energy, internal energy, chemical energy, and light energy. There are other kinds of energy as well, and we shall discuss some of them later in this book. Despite the various forms of energy and the apparent complexity of the idea, two unifying principles underly the concept. The first of these principles is the *conservation of energy*, which we discuss in the next section. The second principle is a relation between mass, energy, and a quantity known as momentum. We discuss momentum in a later section of this chapter and the relation between energy, momentum, and mass in Chapter 10.

7.3 Conservation of energy

The principle of conservation of energy says that although energy can be converted from one form to another, the amount of it is unchanged as time goes on. In other words, energy can neither be created nor destroyed, but is always "conserved," which means that the amount of it stays the same as time goes on.

We illustrate the principle with a few examples.

We first mention *kinetic energy*, which is the energy of motion. If you swing a hammer, you give it kinetic energy, enabling it to drive a nail into a piece of wood. The hammer does work on the nail by exerting a force on it and causing it to move. In the process, the hammer slows down, losing kinetic energy. The lost kinetic energy of the hammer is converted into work done on the nail. The nail encounters a frictional force, which heats up the nail and the wood. The result is that the kinetic energy of the hammer is converted into heat and then into internal energy of the nail and wood.

Potential energy is energy of position or deformation. If you lift up a ball, you give the ball potential energy of an amount that is proportional to its height. If you stretch a rubber band, it gets potential energy due to its deformation.

If you lift a ball and then let it go, it will of course fall. The lower the ball gets, the less potential energy it has, but it moves faster and faster, gaining kinetic energy as it loses potential energy. If we neglect air resistance, we find that the sum of the kinetic energy and potential energy of the ball remains constant as the ball falls. If we include the air, the ball pushes the air out of its way, giving the air a little kinetic energy. Because total energy is conserved, the sum of the kinetic energy and potential energy of the ball slightly decreases.

The instant before the ball touches the ground, its potential energy has been entirely converted to kinetic energy (again, neglecting air resistance). When the ball strikes the ground, the ball momentarily stops moving, losing its kinetic energy. The force of the ground on the ball deforms it from a spherical shape to sort of a pancake shape. This change in the shape of the ball gives it potential energy to make up for the kinetic energy it has lost by stopping. The ball bounces, and its potential energy of deformation gets converted back into kinetic energy. The kinetic energy decreases as the ball rises, as it converts to potential energy of position.

The ball does not bounce quite as high as it was when it was first dropped because it has lost some energy, a little to the air, as we have said, but even more to the floor. The energy given to the air originally is in the form of kinetic energy, but the moving air slows down because of its interaction with the remaining stationary air. Eventually, the air becomes still. The energy of the air does not disappear but is transmitted as *heat energy* (usually just called heat) during the slowing-down process. The energy transmitted to the floor is also transmitted as heat. As a result of the absorption of heat, the temperature of the air and the floor rise a little, and both the air and the floor are said to have more *internal energy*. The temperature of the ball also rises because some of its potential and kinetic energy is converted to internal energy. The higher the temperature of a material, the more internal energy it has.

Chemical energy is energy in matter by virtue of its chemical composition. For example, there is chemical energy in gasoline. When the gasoline burns, it gives off some of this energy as kinetic energy and some as heat. The kinetic energy moves a piston and makes the car go. Chemical energy is one kind of internal energy. We shall understand chemical energy better after we discuss atoms and molecules in later chapters.

Light energy is, as expected, the energy contained in light. As we shall see in Chapters 9 and 11, the energy of light depends both on its intensity (how bright it is) and its wavelength.

Note that in all cases of energy, we do not see the energy itself. Rather, in the case of kinetic energy, we see motion. In the case of potential energy we see position or deformation, and so on. The concept of energy is a theoretical idea, but it has proved extremely valuable in physics.

7.4 Momentum

If an object moves, it not only has kinetic energy but also momentum. For slowly moving objects, the momentum is defined as follows:

The momentum of a body is its mass times its velocity.

Because velocity is a vector, so is momentum. We often give momentum the symbol $\mathbf{p}$. In symbols, momentum is defined by

$\mathbf{p} = m\mathbf{v}$. If the speed of an object is not small (compared to the speed of light), then the definition of momentum is changed, as we shall discuss in Chapter 10.

The importance of momentum arises from Newton's second law of motion, which can be written in the following alternative form: $\mathbf{F} = \Delta\mathbf{p}/\Delta t$, where the symbol Δ means "the change in." Newton's second law in this form says that the force on an object is equal to the change of its momentum divided by the change in time during which the momentum changed, or, in other words, the force is equal to the rate of change of momentum.

It turns out that this form of Newton's second law is more general than the form we have previously given ($\mathbf{F} = m\mathbf{a}$) because the new form holds even when the mass of the object is changing.

As you drive a car, you are burning fuel, and the waste gases leave the car through the exhaust. Therefore, the car loses mass as it moves, so if we want to be very accurate in using Newton's second law to calculate the motion of the car, we need to take account of the change in mass of the car as well as its change in velocity. This change in the mass is accounted for by the change in momentum, which includes both the change in mass and the change in velocity. For the case of a car, the change in mass is a small effect, but in a rocket, the mass of burned fuel expelled can be larger than the mass of the remainder of the rocket. In this latter case, we must use Newton's second law in terms of momentum if we are to calculate the motion at all accurately.

7.5 Conservation of momentum

We saw in the last section that Newton's second law says that the force on an object is equal to the rate of change of its momentum. As a consequence, if the force is zero, the momentum does not change at all.

This is the law of conservation of momentum, which states that the momentum of an object is conserved (does not change as time goes on) unless an external force acts on it.

This law of conservation of momentum is another way of stating Galileo's law of inertia.

Suppose a bomb explodes, and pieces go flying off in all directions. The force that makes the bomb explode is an internal force, not an external force. Suppose the bomb is initially at rest, that is, its momentum is zero. Then, if we neglect the force of gravity compared to the force of the explosive, the sum of the momenta of the pieces immediately after the explosion will be zero as well. Individual fragments will have momenta in different directions, but the total momentum of all the fragments will add up to zero. This can happen because momentum, like force, is a vector, so in adding two momenta, we have to take into account direction as well as magnitude. (If two objects of equal mass travel in opposite directions with equal speed, the total momentum of the two objects taken together is zero. Each object, of course, has momentum different from zero.) After the explosion, the force of gravity cannot be neglected, and the pieces will fall to earth, attaining non-zero momentum on the way down.

7.6 Angular momentum

The earth spins on its axis once a day. Any object that spins is said to have angular momentum. The earth also revolves around the sun once a year. An object that moves in a curved path also has angular momentum. Therefore, the earth has angular momentum by virtue of its spin and by virtue of its rotation around the sun. In order to specify angular momentum of a body one must specify its axis of rotation. For the earth's spin, the axis is a line through the earth's poles that also passes through the center of the earth. For the earth's motion around the sun, the axis is a line through the sun perpendicular to the ecliptic (the plane of the earth's orbit around the sun).

The importance of angular momentum is that, unless there is an external force on a body that acts off the axis of spin or rotation, the angular momentum of the body is conserved. The conservation of angular momentum means that it does not change as time goes on. To a very good approximation, the angular momentum of the earth is conserved. That is why every day has the same duration, and likewise every year. A very small change in the angular momentum of the earth comes from the tidal forces arising from gravity from the moon and sun. Neglecting the forces from other planets, the total angular momentum of the earth, sun, and moon, is conserved.

Chapter 8

Electricity and Magnetism

To study, to finish, to publish.

—Benjamin Franklin (1706–1790)

So far, we have focused on only one fundamental force of nature, gravity. In this chapter, we discuss electrical and magnetic forces, which are now known to be different manifestations of the same force: the "electromagnetic force." As a result of the study of electricity and magnetism, James Clerk Maxwell (1813–1879), a Scottish physicist, discovered that light is intimately related to electromagnetism, in fact, that it is an electromagnetic wave. Because we obtain most of our information about the universe from the electromagnetic waves reaching us from outer space, we need to know something about electricity, magnetism, and light.

8.1 Electricity

Benjamin Franklin did not finish the study of electricity. Rather, he was one of those who pioneered it. He was one of the early people to propose that electric charge comes in two forms, which he called "positive charge" and "negative charge." His definition was arbitrary, as he could have defined the two kinds of charges oppositely.

Franklin discovered that under certain conditions the movement of electric charges in air results in lightning. He also had the notion of the conservation of electric charge. This means that the algebraic sum of all the charges remains constant as time goes on. However, a positive charge may be neutralized by a negative charge of the same magnitude so that the combined charges appear uncharged (electrically neutral). Experiments have revealed that a positive and negative electric charge of the same magnitude can be created or destroyed together, but a single positive or negative charge cannot be created or destroyed. This is the law known as the conservation of electric charge.

According to our understanding of physics, all matter attracts all other matter gravitationally. Electrical forces are different: like charges repel, while unlike charges attract. Here, "like" and "unlike" refer to the sign of the electric charges.

Like the gravitational force, the electrical force between two point-like electric charges is inversely proportional to the square of the distance between them. The force is also proportional to the product of the strength of the charges, which we call Q_1 and Q_2. If we call the constant of proportionality k, then the force law is $F = kQ_1Q_2/r^2$. This force law looks very much like Newton's law of gravitation, with the charges Q_1 and Q_2 taking the place of the masses m_1 and m_2 and the constant k taking the place of the gravitational constant G. Because the charges Q_1Q_2 may be either positive or negative, the product also may be either positive or negative. A positive sign is interpreted as a repulsive force, while a negative sign is interpreted as an attractive force. This is the mathematical statement of the observation that like charges repel and unlike charges attract. But the mathematical law goes beyond that qualitative fact and quantitatively gives the magnitude of the force as well as the direction. The force law between two charges is called Coulomb's law, after the French physicist Charles Coulomb (1736–1806). However, he was not the first to propose this law.

A moving electric charge is called an "electric current." The direction of a current is arbitrarily defined as the direction of motion of positive electric charges. If negative charges are moving, the direction of the current is then opposite to the direction of the motion of the charges. Usually, an electric current is caused by many moving charges. Lightning is caused by a large electric current between two clouds or between a cloud and the earth.

Electric currents can also be made to travel in metal wires, which are called "conductors" of electricity. Some other types of materials do not normally allow current to flow through them, and are called "insulators." Some materials with intermediate properties are called "semiconductors." Most conductors exhibit a resistance to the flow of current, and it takes energy to make the current continue to flow. An exception is a class of conductors known as "superconductors," which have no resistance to the flow of current. Once started, a current in a superconductor continues to flow without any measureable decrease, even without energy input. Some ordinary materials become superconducting at very low temperatures, and so far, no material has been discovered that is superconducting at room temperature. All four types of materials have important applications in modern technology, but to discuss these applications would take us too far from our main subject.

If an electric charge experiences a force on it at some point in space, we say that the force arises because an "electric field" exists at that point in space. The electric field at a point can arise from positive and negative charged particles which are not at the same place as the point. Current flows because of an electric field acting on charges that are free to move when a force acts on them. The electric field is a vector field because it has a direction as well as a magnitude at every point in space and time. The direction of the electric field is defined as the direction of the force on a positive charge placed in the field. Because like charges repel and unlike charges attract, the direction of the electric field is away from a positive charge and toward a negative charge. If there is an electric field in a metal wire, then it is chiefly the negative charges that move, causing the current. Because the moving charges are negative, the motion of the charges is opposite to the direction of the field and the current. In some liquids and gases, both positive and negative charges move. In Chapter 11 we shall discuss the nature of the positive and negative charges.

8.2 Magnetism

The ancient Greeks discovered that certain kinds of stones attract pieces of iron with a force that is called "magnetic." The Chinese may have discovered this even earlier. Magnetic stones contain iron or other magnetic metals such as nickel and cobalt. The fact that iron and a few other metals can have the property of attracting or repelling each other is called magnetism. It turns out that a piece of iron may be magnetized. If the magnetized iron is in the shape of a long, thin rectangular solid, it is called a bar magnet. If one takes two such bar magnets and places them near each other end to end, then the bar magnets either attract each other or repel each other. If they attract, and one bar magnet is turned around so that its other end is near the end of the first magnet, then the magnets repel. It was thus learned that a magnet exerts forces in two different directions, one at each end. The two ends are called a "north pole" and "south pole." The definition of which pole is called a north pole and which a south pole is arbitrary, but to avoid confusion, the definition is the same for everybody. If there is a force on a magnet at some point in space, we say a "magnetic field" exists at that point in space. A magnetic field is a vector field whose direction outside the magnet is defined to be away from a north pole and toward a south pole.

Michael Faraday emphasized the existence of a magnetic field. The field cannot be made visible, but the effect of it can be seen as follows: We put two bar magnets end to end but separated from each other on a piece of paper. Then we sprinkle iron filings (thin slivers of iron) between the bar magnets and find that the filings line up in the direction of the field, as we show in Figure 8.1. These lines are sometimes called field lines and sometimes are called lines of force.

The earth itself has an iron core, and as a result, the earth is a huge magnet. Consequently, the earth has two magnetic poles and a magnetic field. The earth happens to have one of its magnetic poles

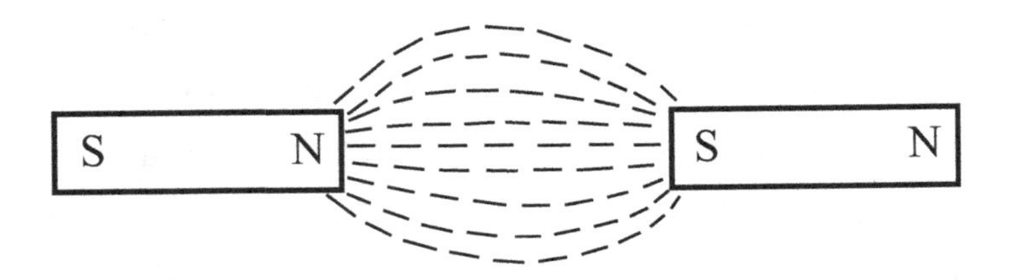

Figure 8.1: Iron filings sprinkled between two bar magnets align themselves along the direction of the magnetic field.

not too far away from the true or geographic north pole of the earth and the other not too far away from the geographic south pole of the earth. The geographic poles of the earth are the points at which the spin axis of the earth intersects the earth's surface. A magnetic compass is a device with an iron needle that is free to rotate, and it rotates so that it lines up with the magnetic field of the earth.

The earth's magnetic field is curved. Michael Faraday imagined lines of field directed from the earth's magnetic pole near its geographic south pole to the magnetic pole near the geographic north pole. The end of the compass needle that points to the earth's magnetic pole in the northern part of the earth is defined to be a north pole. Since unlike poles attract, the earth's magnetic pole near its north georaphic pole is then a south magnetic pole, a fact that may be confusing.

An electric charge of either sign can be isolated, or, in other words, carried arbitrarily far away from a positive charge. If the electric charge is negative, energy is needed to do remove it far from the positive charge. However, one cannot isolate the north and south poles of a magnet. If one cuts a bar magnet in half, one obtains two bar magnets, each of which has a north and a south pole. One can continue to cut the magnets many times with the same result. One can't do this indefinitely, because matter has an atomic structure, as we shall see in Chapter 11. We do not eventually isolate a north pole from a south pole, but rather obtain a smallest magnet, which itself has both a north and south pole.

Thus, there is a profound difference between charge and magnetism: isolated charged particles exist, but, as far as we know, isolated magnetic poles do not. It was Hans Christian Oersted (1777–1851) who first discoverd that an electric current gives rise to magnetism. It is now thought that all magnetism arises from electric currents.

A loop of wire with a current in it gives rise to both a north pole and a south pole, one pole on one side of the plane of the loop and the other pole on the other side of the plane. There are circulating electric currents in all materials because the constituents of materials are moving (spinning) charged particles. However, except under special conditions, the tiny electric currents in most materials spin in many arbitrary directions and thus the current and the magnetism are canceled out. This leaves no observable net current and no observable net magnetism. However, in some materials, such as iron, nickel, and cobalt, currents can be made to circulate in the same direction, giving rise to magnetism on a macroscopic scale. Because the currents are circulating, there is normally no net electric current in any single direction in magnetic materials. We shall say more on the composition of matter in our chapter on atoms.

Not only do moving charged particles give rise to magnetic fields but if a charged particle moves in a magnetic field, the particle will feel a force. This force is quite complicated and depends on the direction the particle is moving. If the charged particle is moving along the lines of a uniform magnetic field (parallel or antiparallel), no force will be exerted on the particle. If a particle moves perpendicular to a uniform magnetic field, the direction of force on the particle is such as to make it move in a circle at constant speed. In other directions, the path of the particle is a combination of a circle and a straight line, called a helix. If a charged particle moves in a nonuniform field, the motion is still more complicated, and the particle may slow down or speed up in addition to moving in a kind of spiral.

8.3 Electromagnetism

Many physicists did experiments on the connection between electricity and magnetism, the greatest of these physicists being Michael Faraday. He found that not only do moving charges give rise to magnetic fields but moving magnetic fields give rise to electric fields. An American physicist, Joseph Henry, independently discovered this effect a little earlier. Electric fields from moving magnets act on charged particles to cause currents. This fact strengthens the connection between electricity and magnetism and suggests that they are two aspects of a fundamental force called the electromagnetic force.

Two important practical devices came out of the work connecting electricity and magnetism: the electric generator and the electric motor. The electric generator is an application of the fact that a magnet moving relative to a loop of wire causes an electric current in the wire. Of course, a source of energy , whether steam, falling water, or something else, is required to get the magnet or loop of wire to move. Thus, an electric generator is a device for converting kinetic energy into electrical energy.

The motor is an application of the fact that an electric current causes a magnetic field that in turn can cause a magnet to move. This motion is the principle behind the electric motor, which is a device for converting electrical energy into kinetic energy. It is outside the scope of this book to go into the engineering details of either the generator or the motor.

James Clerk Maxwell, knowing that electricity and magnetism were related, wrote down equations that mathematically describe the relationship. These equations are the equations of electromagnetism. According to Maxwell's equations, there are regions of space containing no matter but nevertheless contain something. That something is the electromagnetic field. The electromagnetic field can contain stationary electric and magnetic fields and can also contain traveling electromagnetic waves. The traveling waves are simply electric and magnetic fields oscillating perpendicular to each other and perpendicular to the direction of the waves. The equations predict that when electromagnetic waves travel in vacuum, they travel with the speed of light. We therefore obtain the insight that light is simply an electromagnetic wave. We discuss some aspects of light in Chapter 9.

Chapter 9

Wave Motion

> ...the waves we talk about are not just some happy thoughts which we are free to make as we wish, but ideas which must be consistent with all the laws of physics that we know.
>
> —Richard Feynman (1918–1988)

A knowledge of wave motion is essential for understanding both the microscopic world of atomic physics and the universe as a whole. In this chapter we discuss the wave motion of sound and light.

9.1 Sound waves

When one person talks to another, the speaker makes the air vibrate, and the vibrations travel as waves through the air to the ears of the listener. The vibrating air makes parts of the listener's ears vibrate, and signals are sent to the brain, which interprets the vibrations as sound. The "frequency" of vibration, which is defined as the number of vibrations (or cycles) per second, determines the pitch of the sound: the higher the frequency, the higher the pitch perceived by the listener. The ear is sensitive only to a certain range of frequencies.

Frequency is commonly measured in units of number of cycles per second, usually called *hertz*, after Heinrich Hertz (1857–1894), the German physicist who first discovered electromagnetic waves. The normal range of frequencies human beings can hear is from about 16 to about 16,000 hertz. We do not hear vibrations outside that range. Certain animals can hear other frequencies. For example, elephants can hear somewhat lower frequencies and dogs can hear somewhat higher frequencies. A dog can hear and respond to a whistle which makes a tone at a frequency too high for a person to hear.

Even though a sound wave travels from speaker to listener, the air itself does not. The individual molecules of air just vibrate back and forth in the direction of motion of the wave. The molecules hit other molecules and set them in motion, and so the wave travels. Because sound vibrations are back and forth along the direction of motion of the wave, a sound wave is called a "longitudinal" wave.

The distance a wave travels from one maximum compression of the air to the next is called the "wavelength" and is often denoted by the symbol λ. In Figure 9.1 we illustrate the wavelength of a vibrating string.

We can define the frequency of a wave as the number of wavelengths passing a point in one second. This definition is equivalent to our earlier definition. The "period" of a wave is the reciprocal of the frequency. (The reciprocal of any number is one divided by the number.) The period is the time for the wave to travel one wavelength.

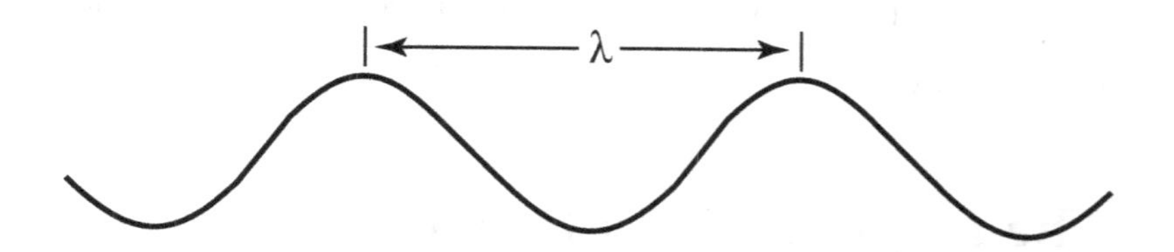

Figure 9.1: The wavelength λ of a vibrating string (or of any other wave) is the distance between successive crests.

We point out a number of characteristics of sound. First, sound moves with a certain speed in the medium, and the speed is characteristic of the medium. In the case of one person talking to another, the medium is normally air, but sound can travel in other media as well, for example, in water or metal. The speed of sound differs in different media. Even in air, the speed depends on the density and temperature of the air, so the speed is different at high altitudes from what it is at sea level. Sound needs a medium to travel; it cannot travel in a vacuum (empty space). The speed of sound in air at sea level under certain standard conditions of temperature and pressure is $v = 344 \quad \mathrm{m/s}$.

There is a relation between the speed of a wave v, its frequency f, and its wavelength λ, namely, that the frequency times the wavelength is equal to the speed, or in symbols $f\lambda = v$. As the wavelength gets shorter, the frequency goes up, and vice versa because the speed remains the same in the medium.

When a small source of sound sets the air vibrating, the sound vibrations do not travel only in one direction, but spread out like an expanding sphere from the source. Furthermore, the sound can bend around objects and be heard behind them. This phenomenon is called "diffraction."

9.2 The Doppler effect for sound

If the source of sound is moving toward the listener, the frequency goes up and is perceived by the brain as a higher pitch. The reason for the frequency increase is that the source is closer to the listener with each successive vibration, so it does not take as long for the successive vibrations to reach the listener. In other words, more vibrations per second reach the ears of the listener. The frequency also goes up if the listener is moving toward the source. When the frequency goes up, the wavelength gets shorter. Conversely, if either the source is moving away from the listener or the listener is moving away from the source, the pereived frequency goes down, and is perceived as a lower pitch. Likewise, the wavelengh gets longer. The change in perceived frequency of a wave arising from motion of the source or the listener is called the "Doppler effect" after Christian Doppler (1803-1853), an Austrian physicist and mathematician.

The Doppler effect can be readily perceived in the siren of an ambulance approaching you. The pitch of the siren is higher than normal as the ambulance approaches, and then, as it passes you and moves away, you can distinctly hear an abrupt lowering of the pitch. The pitch does not depend on how far away the ambulance is but only on how fast it is going.

Of course, the closer the ambulance is to you, the louder it sounds. We say that the sound becomes more intense. Because the the sound spreads out in all directions from the source, its intensity decreases with increasing distance from the source.

9.3 Light waves

In Chapter 8 we saw that electricity and magnetism are different aspects of the same force: the electromagnetic force. Electric forces occur between charged particles, either at rest or in motion. For magnetic forces to arise, charged particles must be in motion. We believe that the forces do not arise because of action at a distance but because of electric and magnetic fields that fill the space between the particles subject to the electric and magnetic forces.

When electric charges not only move but accelerate, for example, vibrate back and forth, the charges give rise to vibrating electric and magnetic fields. These fields travel through space as electromagnetic waves, often called light waves. As we have noted, the existence of these waves is predicted by Maxwell's equations.

In this chapter we treat light only as a wave, but in a later chapter we shall see that light also has certain properties that we associate with particles. The fact that things we usually call waves also have particle-like properties and things we usually call particles also have wave-like properties is incorporated into the theory of quantum mechanics.

There are a number of differences between light waves and sound waves. As we have seen, sound waves travel through a medium, which can be air, water, oil, or something else. In analogy, for a long time it was thought that there must exist a medium for light to travel through. This medium was called the "ether." For many years scientists tried to discover the ether, but all attempts failed. Einstein boldly postulated that the ether does not exist, and that light travels through empty space, called the vacuum. As we have said, sound does not travel through a vacuum.

Another important difference between sound and light waves is that sound waves are longitudinal but light waves are "transverse." By a transverse wave, we mean that the vibrations are perpendicular to the direction the wave is traveling.

The sun emits electromagnetic waves in a large number of frequencies and corresponding wavelengths. Our eyes perceive sunlight as white, but white light is a mixture of many wavelengths. These different wavelengths can be spread out into different positions by a glass prism, which bends light of different wavelengths by different amounts. The different wavelengths appear to our eyes (as interpreted by our brains) as different colors. These colors are the colors of the rainbow, and range from red (the longest wavelength and smallest frequency we can perceive) to violet (the shortest wavelength and largest frequency we can perceive). In Figure 9.2 we show how a prism bends white light into different colors.

The sun emits still longer wavelengths than red, called infrared, but our eyes do not respond to them as light. Likewise, the sun emits shorter wavelengths than violet, called ultraviolet, and again our eyes do not detect them. The range and intensity of wavelengths emitted by the sun (or any other object) is called its "spectrum."

The range of possibilities for the frequency (or wavelength) of electromagnetic radiation is immense. At wavelengths still longer

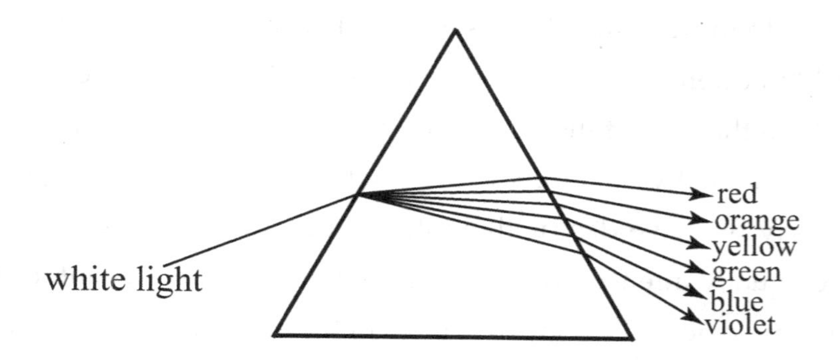

Figure 9.2: Bending of white light by a glass prism into the colors of the rainbow

than those normally called infrared are microwaves and radio waves. At wavelengths shorter than those normally called ultraviolet are X-rays and gamma rays. Although we cannot see electromagnetic radiation outside the so-called "visible spectrum," we can detect such radiation by various detectors. For example, our skin responds to both infrared and ultraviolate light. Infrared light is converted to heat on our skin, and our skin responds to ultraviolet light by tanning. Often we call all wavelengths of elecromagnetic radiation "light," even when we cannot see it with our eyes.

We know light is a wave partly because Maxwell's equations predict waves of electromagnetic fields that radiate when electric charges are accelerated. But we also know that light is a wave because of experiment. If a beam of light is split up into two beams and then recombined and allowd to hit a screen, the intensity of light on the screen is not just the sum of the intensity of the two beams. Rather, in some places the light is more intense than the sum of the intensity of the two light beams, and in other places the combined intensity is less than the sum. This is a typical wave phenomenon, and is called "interference." If the intensity of the light on the screen is greater than the sum of the intensities of the two beams, we say that we have constructive interference, and when the intensity of light is less, we say we have destructive interference. If the two beams of light are equally intense, we can have complete destructive interference in some places. The screen is dark in those places. The fact that light exhibits interference effects shows that light can act as a wave.

9.4 The Doppler effect for light

The Doppler effect exists for any wave motion, and so it exists for light. If a light-emitting source is moving toward us, its frequency is increased and its wavelength is shortened. We call this shift a "blueshift" because the light is shifted toward the blue end of the spectrum. If the light is already blue, the shift will be to the violet or even to the ultraviolet, but we still call the shift a blueshift. On the other hand, if the light source is moving away from us, there will be a shift to longer wavelengths, called a "redshift" in the light. Because of these shifts, if we know the intrinsic wavelengths (or frequencies) of emitted light, we can tell whether an object is moving toward us or away from us. We can also tell how fast the object is moving because the faster the source moves, the greater the shift in wavelength. The formula that is used to calculate the Doppler effect for light is not exactly the same as the formula for sound because light moves at relativistic speed — in fact, the ultimate relativistic speed. However, as we do not give the formulas in any case, the difference between light and sound is not important for us.

The Doppler effect for light is an important tool to enable us to tell about the motion of the stars and galaxies. The stars even in our own galaxy, the Milky Way, are too far away from us to enable us to perceive their motion by noting their change in position in a short period of time. Their apparent change in position is primarily the effect of the rotation of the earth. The revolution of the earth around the sun and the motion of the sun in our galaxy are secondary effects.

However, we can learn about the motion of distant stars or other light-producing objects such as galaxies from the spectra of light they emit. If a star or galaxy is moving toward us, then the wavelength of the light we perceive is shifted toward shorter wavelengths, or, in our shorthand notation, we perceive a blueshift. If the object is moving away from us, then the light is shifted toward longer wavelengths, and we perceive a redshift. We shall discuss the use of the Doppler effect for cosmology in Chapter 19.

Chapter 10

Relativity

God is subtle, but he is not malicious.

—Albert Einstein (1879–1955)

Henceforth space by itself, and time by itself, are doomed to fade away into mere shadows, and only a kind of union of the two will preserve an independent reality.

—Herman Minkowski (1864–1909)

10.1 Limitations of Newton's laws

Newton's laws of motion and his law of gravitation survived intact from the seventeenth century, when they were formulated, until the twentieth century, when they were superseded. It is not that Newton's laws were proved to be totally wrong but that the laws have a smaller range of validity than previously thought. For fast-moving objects, Albert Einstein's special theory of relativity, proposed in 1905, must be used, and for objects with very large masses, Einstein's general theory of relativity, proposed in 1915, must be used.

In the microscopic domain, quantum mechanics must be used. In this chapter we confine ourselves to the modifications required when objects move very fast or have very large masses.

Newton's laws are based on a conception of space and time which is in accord with our intuition, but intuition is not always a reliable guide to the way nature behaves. According to Newton, space is three-dimensional and flat. By a flat space we mean a space in which parallel lines never meet and in which the sum of the angles of a triangle is 180 degrees. Furthermore, according to Newton, all observers measuring the size of an object will obtain the same result. Also, Newton's concept of time is that it is the same for everybody. Every observer measuring the time between two events will obtain the same result, within the accuracy of his clock.

What goes wrong with Newton's notions of space and time is that he does not take into account that the speed of light in a vacuum (in empty space) is a limiting speed that cannot be exceeded by any object, as experiments have demonstrated. The speed of light in vacuum is normally denoted by the letter c, and its value is given approximately by $c = 300,000$ kilometers per second (km/s) (about 186,000 mi/s). It is one of the fundamental constants of nature.

The speed of light is such a large number that it is easy to overlook effects which become important only at speeds approaching the speed of light. To help us gain a feeling for how great the speed of light is, it is convenient to introduce units of distance that tell how far light travels in a given amount of time. For example, a light-second is the distance light travels in a second. That distance is more than seven times around the earth at the equator. The distance from the earth to the moon is a little over a light-second. The distance from the earth to the sun is much larger: about 8 light-minutes. The sun and moon appear the same size to the eye only because the moon is so much closer than the sun. A distance unit sometimes used in astronomy is the light-year, the distance light travels in a year. The nearest star to the earth is about 4 light-years away. A more common unit of distance in astronomy is the "parsec," equal to 3.26 light-years, but we shall use the light-year. Many stars are comparable in size to the sun, and some are much bigger, but they appear as points of light in the sky because of their huge distances from the earth.

10.2 New laws of motion

As we have said, when objects move at speeds that are not small compared to the speed of light, Newton's laws break down, and Einstein's special theory of relativity must be used. Special relativity rests on two postulates:

The speed of light in vacuum is the same constant when measured by all observers in inertial frames of reference.

The laws of nature are the same for all observers in inertial frames of reference.

(An inertial frame of reference is defined in Chapter 5.)

The first postulate is astonishing, as it is completely contrary to what happens for objects going at small speeds. For example, consider a boat going downstream in a river. The boat travels at a certain speed v in the water, which is flowing at another speed v_r. Then the speed of the boat v_o relative to an observer on shore will be greater than its speed relative to the water. In fact, the speed relative to the shore will be the sum of the two speeds: $v_o = v + v_r$. If the boat instead travels upstream, the speed v_o relative to the observer on shore is the difference of the speeds. This relation is known as the nonrelativistic law of addition of velocities. It is the law that is compatible with Newton's laws of motion.

Newton's laws are often called "nonrelativistic" because they are incompatible with Einstein's special relativity. The word nonrelativistic is a little misleading because Newton's laws do incorporate a kind of relativity in that absolute motion at constant velocity cannot be measured. Only motion that is relative to something else is measurable. The relativity of Newton's laws is sometimes called "Galilean relativity" to distinguish it from "Einstein relativity." (It was Galileo who first proposed that absolute motion at constant velocity could not be measured, only motion relative to something else.) But we will follow the rule that physicists ordinarily use and call a law nonrelativistic if it is compatible with Galilean relativity rather than with Einstein's special relativity.

We return to the girl on the boat. If she shines a flashlight either in the direction the boat is moving or in the opposite direction, the speed of light coming from the flashlight will be the same constant c as measured by the girl or by an observer on shore. According to special relativity, no signal can travel faster than the speed of light in vacuum. (The speed of light in air is only a little slower than the speed of light in a vacuum, and we neglect the difference for our purposes.)

If the speed of light were c only in one reference frame, that frame would be singled out from all other frames, in contrast to the idea that the laws of nature are the same in all inertial frames.

So the nonrelativistic law of addition of velocities breaks down when one of the velocities is c. The breakdown is not sudden, but as either one of the two velocities involved becomes large, the deviation from the nonrelativistic law of addition of velocities increases. The new law of combination is more complicated than the nonrelativistic additive law, and we do not write it down here.

We are used to observing objects going much slower than the speed of light. The speed of light itself is so great that we cannot notice without a careful measurement with sophisticated instruments that light speed is a constant independent of the motion of the source or observer (so long as the observer is in an inertial frame). For this reason, our intuition fails us when we are concerned with the properties of light and fast-moving objects.

But the failure of the law of the addition of velocities is just one of the profound consequences of the special theory of relativity. The very fabric of space and time, as assumed by Newton, is altered.

Time is no longer the same for everyone. Time is something measured on clocks, and not all clocks run at the same rate. It is not that clocks keep bad time but rather that clocks moving at different speeds through space keep different times. The faster a clock travels in space, the slower it runs in time. The speed of light is a limiting speed, and no object with mass greater than zero can move as fast as light. If a clock could move with the speed of light, it would appear as if the clock did not run at all. Thus, the theory says that time measured on a clock moving with the speed of light would stop altogether. (The theory also says that the clock could not move as fast as the speed of light. We shall discuss this later in this chapter.)

Thus, there is not a single time in the universe, but many times, each measured with a different clock. The faster the clock moves in space, the slower it moves in time. In Figure 10.1 we illustrate the fact that a moving clock runs slow. The figure shows only an example; if the clock moves faster to the right than the speed v shown in the figure, the clock would read earlier than 12:25, while if the clock moves slower, it would read later than 12:25.

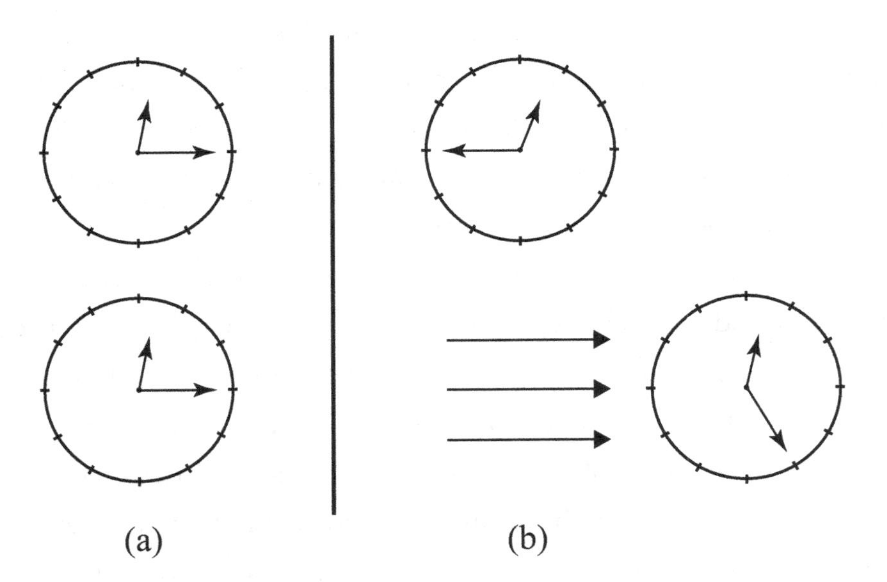

Figure 10.1: If two clocks read the same time when they are together (in (a), 12:15), and if one clock moves sufficiently fast while the other remains stationary, then the moving clock shows a shorter elapsed time (in (b), when the stationary clock reads 12:45, the moving clock reads only 12:25).

Another consequence of special relativity is that the length of a moving object becomes shorter in the direction of motion. As the object moves faster and faster, its length contracts more and more, and if it could move at the speed of light, its length would contract to zero. In Figure 10.2 we illustrate the fact that a moving object contracts in the direction of motion.

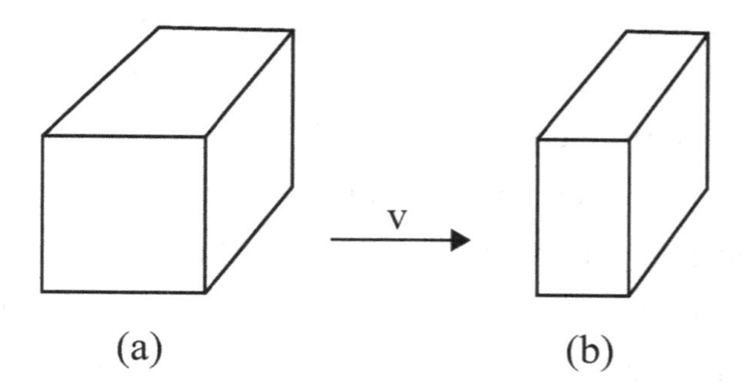

Figure 10.2: Illustration that a moving object contracts in the direction of motion. In (a) we show an object when at rest, and in (b) we show the same object in rapid motion to the right. Note that the object does not contract in directions perpendicular to the motion.

We have already noted that the nearest star to the sun is about four light-years away. This means that it would take about four years for a beam of light from the sun (or earth) to reach the nearest star. However, if a person could travel in a space ship from the earth to the star at nearly the speed of light, the traveler would arrive much sooner than four years according to his own clock. How is that possible? The traveler would find that the distance to the star shrinks so that the star would not be nearly as far off. To an observer on earth, it would seem as if it took the traveler a little more than four years, but the observer would see the traveler's clock slow down and would realize that according to the traveler's clock the time would be less than four years. The traveler on the space ship thinks his clock is moving normally but that the distance to the star shortens, whereas the person on earth thinks the distance remains the same but the traveler's clock slows down.

Who is "right," the traveler or the observer on earth? The answer is that both are right according to their own measurements, which do not need to agree because the two observers are making measurements from different reference frames.

According to Newtonian mechanics, two different observers always measure the same distance in space between two separated events. However, the equations of special relativity mix space and time. What this means is that if one observer sees two events occurring in different places at the same time, another observer in a different inertial frame will observe the two events taking place at *different* times. A generalization has to be made to the concept of distance. The new "distance" between two events is the distance, not in three-dimensional space but a distance in a four-dimensional spacetime. This four-dimensional distance, appropriately defined, is the same for all observers in different inertial frames. (Recall that time can be treated like a distance by multiplying it by the speed of light.)

Although the theory of the relativity of space and time was invented by Einstein, it was Herman Minkowski (1864–1909), a Russian mathematician who did most of his work in Germany and Switzerland, who developed a description of relativity that treated space and time on an equal footing. Minkowski's formulation of special relativity was helpful to Einstin in developing the general theory of relativity.

Still another consequence of special relativity is the equivalence of mass and energy. The famous Einstein equation relating the mass m of an object at rest and its energy E is $E = mc^2$, where c is the speed of light. Furthermore, if an object is set in motion, its inertia increases. Some physicists say that the mass increases with speed, as mass is a quantitative measure of inertia. It is more convenient, however, to define the mass of an object to be its mass at rest, and to say that the inertia is proportional to the energy.

The momentum of a fast-moving object is no-longer just its mass multiplied by its velocity but is greater. We omit giving the formula, which contains the mass and velocity, as well as the speed of light. If an object is moving, the energy depends on the momentum of the object as well as on its mass. The general formula is a little complicated, and so we do not give it. For an object at rest, the formula relating mass and energy becomes $E = mc^2$. If the object has no mass at all, it can never be at rest, but it moves with the speed of light. In this extreme case, the energy of the object is proportional to its momentum p, the constant of proportionality being the speed of light. In symbols, for an object without mass, the general formula becomes $E = pc$. If the object has a mass greater than zero, it cannot be made to go as fast as light because the special theory says that it would take an infinite amount of energy to make the object move that fast.

According to Maxwell's theory of electromagnetism, light is predicted to travel at the same constant speed in vacuum in all inertial frames of reference. One might suppose, therefore, that unlike Newton's equations of motion, Maxwell's equations are compatible with the theory of special relativity. That supposition is correct, as has been demonstrated by direct calculation.

10.3 The twin paradox

The prediction that moving clocks run slow has been amply demonstrated by experiment. Convincing evidence comes from the realm of subatomic physics, but we shall defer discussing that evidence until after we discuss atoms and elementary particles. Here we shall discuss evidence provided by extremely accurate clocks. Three such clocks were compared at rest and found to keep the same time within experimental error. Then one clock was kept at rest, while the other two clocks were put on airplanes and flown around the world, one in each direction.

The earth rotates west-to-east, as we can tell from the apparent daily movement of the sun east-to-west. If an airplane flies west-to-east, it is flying in the same direction as the earth is rotating, and so its speed is greater than the speed of rotation of the earth. Consequently, the clock on that airplane moves at a slower speed than the clock that is stationary with respect to the earth. Therefore, when the two clocks are compared after the trip, the clock on the airplane is slightly behind the clock that has remained stationary on the earth. On the other hand, the airplane traveling east-to-west moves slower than than the earth, so its clock runs slightly ahead of the stationary clock.

The experiment confirmed these predictions. The differences among the clocks were found to be only a tiny fraction of a second, owing to the fact that the speed of the airplane (about 900 km/hr) is only a small fraction of the speed of light.

According to special relativity, it is only relative speeds that count. Why then, was a difference found between the clocks in the two airplanes, when both moved at the same speed relative to the clock that was stationary on the earth? The answer is that because the earth rotates, an observer on the earth is not strictly speaking in an inertial frame, where the predictions of special relativity are valid.

Another complication arises from the earth's gravity. Although for most purposes, the earth's gravity is sufficiently weak that Newton's law of gravity is adequate, for very precise measurements it is not. In such situations, Einstein's general theory of relativity, which is a better theory of gravity than Newton's, must be used. According to general relativity, clocks are affected by gravity. Because the clocks in the airplanes are not at the same height as the the clock on earth, the gravity they feel is different and they do not run at the same speed as the clock on earth. After correcting for the effect of gravity on the clocks and the fact that the earth is not strictly speaking an inertial frame, the experimenters obtained agreement with the prediction of special relativity that clocks moving in space slow down in their measurement of time.

We now turn to the "twin paradox," as it illustrates the importance of making measurements from an inertial frame. The twin paradox is not really a paradox, but appears to be one to those who do not understand the restriction that the special theory of relativity does not necessarily give the right answer unless measurements are made from an inertial frame. The statement goes as follows:

There are twin boys on earth, each ten years old. One boy goes off in a space ship at nearly the speed of light. (As a practical matter, there is no way for a space ship to achieve this speed, but we are doing a thought experiment.) The boy remaining on earth observes that the clock on the space ship runs slow. Not only does the clock run slow but so do the heartbeats and aging processes of the boy in the space ship. But the boy in the space ship observes that the clock of his twin on earth runs slow. After a time the boy in the space ship returns to earth. Each boy expects to be older than his twin, but each cannot be older than the other. What actually happens?

To simplify the problem, let's neglect the fact that the earth is not strictly in an inertial frame. The difference is small. Because the twin remaining on earth remains for all practical purposes in an inertial frame during the entire journey of his brother, his observations are in accord with the predictions of special relativity. Therefore, his conclusion is correct that he will be older than his brother after his brother returns to earth. However, the twin in the space ship cannot remain in an inertial frame and get back to earth. In order to return, the space ship must turn around. While it is turning around it is accelerating. (Remember that a change in direction is an acceleration.) While one is in an accelerating frame one cannot conclude that moving clocks run slow. It turns out that the twin in the space ship will observe that, while he is turning around, the twin on earth will age rapidly. After turning around, while heading back to earth, the twin remaining on earth will appear to age less rapidly than the twin in the space ship, but the effect will not be large enough to undo the rapid aging during the acceleration. So when the space ship gets back to earth, and the twins look at each other, both will observe that the twin remaining on earth is older. If the speed is fast enough and the trip lasts long enough, the twin on earth can be an old man while the traveling twin is still about ten years old.

10.4 Light as a limiting speed

The special theory of relativity says that no object with mass can be accelerated to the speed of light in vacuum, as it would take an infinite amount of energy make it achieve this speed. However, what about the possibility of sending a signal from one observer to another at a speed faster than the speed of light? Is this possible? The postulates of special relativity do not enable us to answer this question directly. However, we can answer the question by making use of the principle of causality.

According to special relativity, two events separated in space that appear to be simultaneous in one inertial frame are not simultaneous in another inertial frame moving with respect to the first. Consider a signal sent from one person to another in a different location. Suppose two observers in different inertial frames see the signal emitted at the same time by the first person. Then each observer will observe the signal arrive at the second observer at a different time. However, if the signal is transmitted at the speed of light or slower, then both observers can agree on one thing: namely, that the signal will arrive at the second person *after* it has been sent by the first. This temporal ordering is what we mean by causality. To no observer will the signal seem to be received by the second person before it is sent by the first, or, what amounts to the same thing, no person will say that the signal was sent by the second person and travels to the first.

However, according to the equations of special relativity, if a signal could be sent faster than the speed of light, then to an observer in some inertial frame, causality would be violated, with the signal appearing to go the other way. A violation of causality can lead to seemingly absurd conclusions, or, in other words, to lead to conclusions which appear to be self contradictory. For example, one could in principle go backward in time and kill ones mother before being born.

We conclude that special relativity and causality together say that no signal can be sent faster than the speed of light in vacuum.

10.5 The equivalence principle

Special relativity is special in that it applies only to observations made from inertial frames. Einstein generalized the theory to observations made from accelerating frames. He called his theory the general theory of relativity, and it is a theory of gravity differing from Newton's theory, although it reduces to Newton's theory in the domain in which Newton's theory is correct.

One of the main pillars on which the general theory is built is the "principle of equivalence." Einstein was impressed by the fact, first stated by Galileo, that all objects a little above the surface of the earth fall with the same acceleration (neglecting air resistance). Einstein was led to consider another situation in which there is no gravity but the effects are equivalent to those of gravity. Imagine being in a closed elevator far from any gravity and that the elevator is being accelerated upward with constant acceleration. Suppose you take a large stone and a small stone and let them go. They will not fall (because of the absence of gravity), but the elevator floor will accelerate upward to meet them. It will seem to you in the elevator as if the stones fall because of gravity, and they will hit the floor of the elevator at the same time. Einstein turned this idea into a principle: the principle of equivalence, which says

> There is no way locally to distinguish beween motion in an inertial frame with a force of gravity and motion without gravity in an acclerating frame.

This statement is called "the principle of equivalence." It follows from this principle that all objects subject to gravity, whether heavy or light, will fall with the same acceleration.

10.6 Gravity as curved spacetime

The theory of general relativity has built into it the principle of equivalence, but the theory is much more than that. It is a theory in which the presence of matter distorts the very fabric of spacetime. A body subject to gravity moves in a curved path because space and time are curved by the matter that is responsible for the gravity.

Consider, for example, a body constrained to move on the surface of a sphere. Its path will always be curved because the surface is curved. Space in the presence of matter is curved, so bodies in that space are constrained to move on curved paths. The curvature in any place and time is what it is, so all bodies starting with the same velocity in the same place at the same time will follow the same path.

Near a large mass, the curvature of space is like a hole, and all objects will fall into the hole with the same acceleration. This is Einstein's explanation of why a small stone and a large stone dropped from a building follow the same path with the same acceleration. Because the motion is "natural," we do not feel any force while we are in free fall. As we have noted, when somebody jumps off the roof of a building, it is not the fall that hurts; it is the sudden stop.

There is a caveat. If a body is large, not all of it is in the same place, and different parts of the body can be in places where space has different curvatures. If so, there will be stresses, called tidal forces, that tend to rip the body apart.

Tidal forces occur in Newton's theory of gravity as well as in Einstein's because the force of gravity on a large object is not necessarily the same on all parts of the object. Consider the ocean tides on earth. When the moon is at a certain place over the ocean, the part of the ocean nearest the moon will experience greater acceleration than other parts of the ocean. This will cause part of the ocean to bulge up in the direction of the moon. If the bulging part is near the shore, we experience a high tide. (The sun also influences the tides, but less than the moon because tidal forces decline with distance more rapidly than an inverse square, as can be calculated mathematically.) Newton's theory of gravity accounts for the ocean tides on earth just as well as general relativity because the tidal forces are not large enough to make any difference in the predictions of the two theories. But in situations with much larger tidal effects, i.e., much larger changes in curvature in a small region of space, the two theories have different consequences. The theories will not give identical predictions near the surface of a very dense star, for example, near a neutron star (which we shall discuss in Chapter 18).

So there are differences between the effects of gravity and the effects of being in an accelerating elevator, but the effects are indistinguishable in a small enough region. For example, on earth gravity is toward the center of the earth, which is in entirely different directions on different parts of the earth. But the force on two bodies sufficiently close together, the force of gravity will be indistinguishable from the apparent force arising from an accelerating elevator. That is why we say that "locally" gravity and uniform acceleration are indistinguishable.

Not only is space curved by gravity but clocks are slowed down. The stronger gravity is, the more clocks will be slowed. An observer in the rest system of the clock will not notice the clock running slow because the observer's body and senses will also slow down. The clock slows down as observed by an outside observer far from the region of strong gravity.

Einstein himself made two predictions based on his general theory. The first is that the orbit of the planet Mercury should not be an exact ellipse but should be an approximate ellipse whose distance of closest approach to the sun (the perihelion) should be in a different place with each revolution. We call this shift a precession. A small amount of precession should occur because of perturbations from other planets but there should remain a small perturbation arising from general relativity itself. This prediction accounted for a previously unexplained precession of Mercury's orbit. The orbits of other planets should also deviate from ellipses, but the effect is largest for

Mercury because it is closest to the sun. In Chapter 2 we discuss the planets, including Mercury, in more detail.

The second prediction is that light from a star passing near to the sun on its way to earth should be bent by the sun's gravity. Normally, a star nearly behind the sun cannot be seen because of the intense light from the sun itself. However, during an eclipse of the sun, light from stars near the sun can be observed and the amount the light bends shows up as an apparent shift in the position of the star compared to where it is observed at night. In Figure 10.3 we illustrate the bending of light during a solar eclipse.

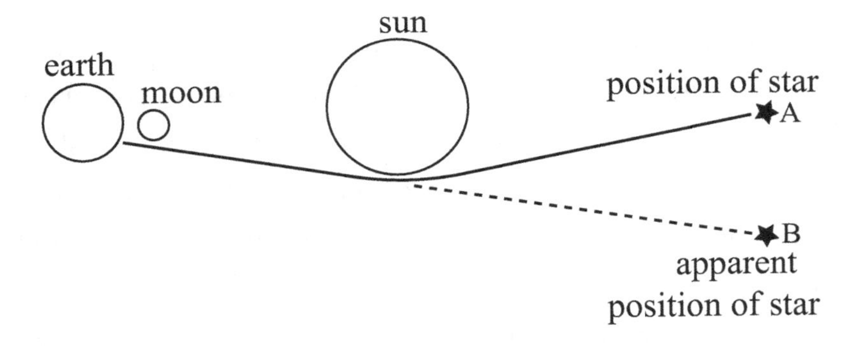

Figure 10.3: A star at position A apparently moves closer to the star at position B because of the bending of light by the sun's gravity. The sizes and positions of the earth, moon, sun, and stars are not to scale, and the amount of the bending of the light is exaggerated.

10.7 Black holes

General relativity allows for gravity to become so strong near a large mass that nothing can escape from the region, not even light. Such a region is called a "black hole." In Einstein's explanation, spacetime is so curved near a black hole that inside a certain sphere surrounding it, both matter and light are trapped by the black hole and can only fall into it.

There is a simple, but not quite correct, way to understand how gravity can be strong enough in a region to prevent anything from escaping from it. We begin by considering throwing an object straight up at the surface of the earth. If you throw a ball at a low speed, it will not get very high before it stops and falls back to earth. If you throw it faster, it will go higher before it returns. If you shoot a rocket fast enough, it will go many kilometers high before stopping and returning to earth. There is a certain speed, called the "escape speed" or "escape velocity," such that if you shoot the rocket with that speed or greater, it will escape the earth's gravity and never return to earth. The escape velocity at the surface of the earth is about 40,000 km/hr. If the earth had more mass, it would exert a larger gravitational force, and the escape velocity would be higher. Now think of a body with so much mass that the escape velocity is larger than the speed of light. Then no object, not even a beam of light, could escape from it. This argument just makes plausible that a black hole can exist. The correct argument makes use of the general theory of relativity.

Around a normal black hole there is a spherical shell, called the "horizon." The shell is not a physical object but consists of empty space. The area of the horizon depends on the mass of the black hole—the greater the mass, the greater the area of the shell. Any object that penetrates the horizon cannot escape, and will continue to fall inward. If you fall across a horizon of a black hole, you will not notice anything special there (if tidal forces are sufficiently small), but you will never be able to cross the horizon again in the opposite direction. According to general relativity, the mass of a black hole collapses to a point at the center, but general relativity probably breaks down under such extreme conditions, especially because quantum effects (to be discussed later) become important.

There is good evidence for the existence of black holes in our universe, but we shall defer discussing this evidence until we have treated quantum mechanics, elementary particles, and stars.

Chapter 11

Atoms

> According to convention, there is a sweet and a bitter, a hot and a cold, and according to convention there is color. In truth there are atoms and the void.
>
> —Democritus (c. 460–357 BCE)
>
> Everything is made of atoms.
>
> —Richard Feynman (1918–1988)

In our modern view of nature, the laws governing the behavior of the smallest bits of matter are related to the evolution of the universe as a whole. Contrary to the early view of atoms, what we now call atoms are not the smallest things there are but are made up of still smaller stuff. Also, atoms can be broken apart into smaller constituents.

11.1 Early ideas on atoms

The Greek philosophers speculated on many things concerning the physical world. They often hypothesized about the way nature is, but before Aristotle, seldom made any attempt to verify their hypotheses. For that reason, the idea that ordinary matter is made of small things that cannot be subdivided (atoms) existed alongside the idea that matter can be infinitely subdivided.

An early Greek who postulated the existence of atoms was Anaxagoras (500–428 BCE). His ideas were sharpened by Democritus (c. 460–357 BCE), who said that matter is made of different kinds of invisible atoms existing in otherwise empty space. As far as we can tell, Democritus made no attempt to think of a way to determine whether atoms exist or not. Evidently, the idea of atoms just appealed to him philosophically.

The idea that matter could be infinitely subdivided appealed to other Greek philosophers, among them Aristotle. Although Aristotle emphasized observation, he did not restrict his theoretical ideas to objects he could observe, and even in cases where he did observe nature he often drew the wrong conclusions from his observations. Aristotle was a great philosopher but his science left much to be desired.

Aristotle was unlucky in much of his science. Many theories go beyond observation. Aristotle must have seen that he could subdivide matter into smaller and smaller pieces. He could not be expected to know that at a scale far too small for him to observe, matter could not be subdivided further without undergoing a profound change in its properties.

11.2 Evidence for the existence of atoms

Indirect evidence that matter is composed of atoms was suggested more than 200 years ago. One piece of evidence was that definite substances could be recovered from other apparently homogenous substances, as if the seemingly homogeneous substances consisted of small individual bits of other things.

Another piece of evidence came from the behavior of gases. A gas in an enclosed container exerts a pressure on the walls of the container. (In Chapter 5 we defined pressure as force per unit area.) This pressure increases as the temperature of the gas increases and as the amount of gas in the container increases. It is as if the gas is made of many particles moving at random in the container. At any given moment some of the particles hit the walls and exert a force and a pressure on the walls. If the amount of gas is increased, more particles hit a wall per unit time, and so exert more pressure. If the temperature is increased, the particles of the gas move faster, and so exert more pressure. The faster the particles move, the more kinetic energy they have. Not all the particles of the gas have the same kinetic energy, but one can talk about the average kinetic energy of the particles. The temperature is proportional to this average kinetic energy.

While we are on the subject of temperature, we recall that we have discussed two different temperature scales, Fahrenheit and Celsius. In the Fahrenheit scale, water freezes at 32° and boils at 212°, while in the Celsius scale, water freezes at 0° and boils at 100°. The Celsius scale is used throughout most of the world except for the United States. There is a third scale often used in science, and that is the "absolute" or Kelvin scale, after Lord Kelvin (1824–1907), a British physicist born in Ireland as William Thomson. The scientific law that at constant pressure, the volume of an enclosed gas is proportional to the temperature, holds when the temperature is given in the Kelvin scale. Also, the statement we have given in the previous paragraph that the temperature is proportional to the average kinetic energy holds when the temperature is given in kelvins, abbreviated K. The coldest temperature is 0 K, although there is a law that although we can get very close to 0 K, we cannot in fact achieve it. (The ° is customarily omitted in kelvins.) In kelvins, the freezing point of water is 273 and the boiling point is 373. Note that there are 100 degrees difference between the boiling point and freezing point of water in both the Kelvin and Celsius scales. Thus, a one degree difference in both these scales is the same, and the scales differ only in a displacement of scale by a constant value of 273 degrees (0 K is the same as -273° C).

It was discovered that many substances, called "compounds," are made of simpler substances, called "elements," which combine in definite proportions. The smallest amount of a compound is called a molecule, and the smallest amount of an element is called an atom. For example, water is a compound, made by combining two elements, hydrogen and oxygen, in definite proportions. By weight, there are eight parts oxygen for every part hydrogen. It was further discovered that there exists a smallest amount of water, a single molecule. This molecule is made of two atoms of hydrogen and one atom of oxygen. The reason that the oxygen in water weighs eight times as much as the hydrogren is that a single atom of oxygen weighs 16 times as much as a single atom of hydrogen. The number of each kind of atom combining into a molecule is exact, but the fact that an atom of oxygen is 16 times as heavy as an atom of hydrogen is not exact, for a reason we discuss in Chapter 15.

There are more than 100 known elements, some of which are unstable and do not ordinarily exist on the earth but decay in a process known as "radioactivity" into other elements. We discuss radioactivity later in this chapter. Most of the known unstable elements have been made artificially, as we shall see in Chapter 15. The lightest stable element is hydrogen. The heaviest element naturally occuring on earth in sizeable amounts is uranium, which is unstable but is so slow to decay that a good fraction of the amount that was on the earth when it was formed still exists.

An ancient pursuit of obscure origin was called alchemy, and had as one of its principal aims, the transmutation of so-called base elements, like lead, into gold. Isaac Newton was one of the greatest physicists who ever lived, but he devoted years of his life to alchemy. The alchemy of the ancient and medieval periods and of Newton really had no chance to succeed because the experiments were chemical. We now know (see Chapter 15) that the nature of a chemical element is determined by the nuclei of its atoms, and so any transmutation of the elements must involve nuclear physics. The existence of atomic nuclei was not known until early in the 20th century.

Strong evidence for the existence of molecules came from interpretation of the phenomenon of Brownian motion. This is the random motion of small dust particles in air or in a liquid. In 1905 Einstein calculated that if dust particles were struck randomly by molecules, they would undergo just the sort of motion that was observed. The motion occurs because the dust particles are so small that at any given moment, they are struck by more molecules on one side than on the others. At a later moment, they are struck more on another side and so change their direction of motion. Many scientists who had previously resisted the notion of atoms because they could not be observed (with the microscopes available to them), were convinced by Einstein's explanation of Brownian motion. Now we have special microscopes capable of enabling us to see single atoms.

A molecule may consist of two or more different kinds of atoms or may consist of one or more of the same kind of atom. For example, the two principal gases of the air of our atmosphere are nitrogen and oxygen. Each molecule of gaseous nitrogen is composed of two atoms of nitrogen bound together by electrical forces, and likewise for gaseous oxygen. A rarer gas in our atmosphere is carbon dioxide, a molecule of which contains one atom of carbon and two atoms of oxygen. On the other hand, a molecule of helium gas contains only one atom of helium. Atoms are bound together to form molecules by electrical forces.

11.3 The composition of atoms

In the last section we have stated that molecules, except in the case of single-atom molecules, consist of two or more atoms bound together by electrical forces. But atoms are electrically neutral, so why should electrical forces exist between them? The answer is that although atoms are neutral as a whole, they consist of smaller particles that are electrically charged. If two atoms are far apart compared to the size of the atoms, each atom acts like an electrically neutral object because the effects of the positive and negative charges cancel each other out. However, if the atoms are close together, each feels the electrical forces from the charged constituents of the other because the positive and negative charges are not in exactly the same places.

It was further discovered that under certain circumstances, atoms could have a net positive or negative electric charge. Such atoms are called positive or negative "ions." The observation of ions showed that atoms were not elementary because charged bits could be added to them or removed from them.

What is the nature of these electrically charged bits? Sir J. J. Thomson (1856–1940), an English physicist, and colleagues, working in the last few years of the 19th century, discovered that the negatively-charged bits are tiny particles, each with the same ratio of charge to mass. The experiment consists of a tube from which some of the air is evacuated and which has metal at each end. If a voltage is applied between the ends, charged particles are emitted from the negative end, called the "cathode," and travel to the positive end, called the "anode." The charged particles emit electromagnetic radiation (light) when striking the residual air molecules in the tube, thereby enabling the path of the particles to be seen.

Thomson showed that that the particles were deflected by a magnet, thereby proving that the particles were electrically charged. The direction of deflection showed that the charges were negative. By measuring the amount of deflection caused by the known magnetic field of the magnet and by measuring the strength of a transverse electric field required to eliminate the deflection, Thomson was able to determine the ratio of the electric charge e to the particle mass m.

He found that all the particles had the same ratio of e/m, as if all of them were identical. The particle is called the electron, and was discovered by Thomson in 1897. It is the first particle discovered that we still believe to be elementary, and the only one discovered in the 19th century.

After the discovery of the electron, it was proposed that a positive ion is an atom from which one or more electrons are removed, and a negative ion is an atom to which one or more additional electrons are attached. This is still the picture we have today.

11.4 The atomic nucleus

Thomson proposed a model of the atom, sometimes called the "plum-pudding" model. According to this model, the positive charge is spread out fairly uniformly over the volume of the atom (the pudding), and the negative charge, consisting of electrons (the plums), are distributed within the positively charged pudding. This picture did not last long because of work done from 1906 to 1909 by Ernest Rutherford (1871–1937) and his assistants, Geiger and Marsden. Rutherford was born in New Zealand, but did his most important work in England.

It had been known for several years that some kinds of substances are "radioactive," which means that these substances emit rays of various kinds. The three kinds of rays usually emitted in radioactivity are called α, β, and γ (alpha, beta, and gamma), and they were known to be positively charged, negatively charged, and electrically neutral respectively. The negatively charged rays were later shown to be electrons and the positively charged rays were shown to be doubly-charged particles. The electrically neutral rays were later shown to be very energetic electromagnetic radiation.

Rutherford's assistants let α particles from a radioactive source strike a thin gold foil, and they observed how the α particles were scattered from the atoms of the foil. The results were surprising. According to the Thomson model, the positive charge on the gold atom is spread out and not concentrated enough to exert a strong repulsive force on the α particles. Therefore, it was expected that the the α particles should be scattered only through very small angles. Instead, some of the α particles were scattered through large angles, as would be the case if they had come very close to a charge concentrated in a region that is much smaller than the atom as a whole.

Rutherford explained the scattering with a model of the atom which is very different from the Thomson model. According to Rutherford's model, nearly all the mass of the atom is concentrated in a tiny positively-charged nucleus, while the electrons orbit around the nucleus much like the planets orbit around the sun. Of course, an atom is on a tiny scale compared to the solar system. We now know that the radius of an atom is much less than a billionth of a meter (actually, about 10^{-10} m), and the radius of a nucleus is up to 100,000 times smaller still. In Rutherford's nuclear model, most of an atom consists of empty space. In Figure 11.1 we show a schematic drawing of Rutherford's model of an atom.

Our present picture of the atom, while not identical to Rutherford's model, is similar in that the electrons are relatively far away from the central nucleus. We now know that radioactive emissions

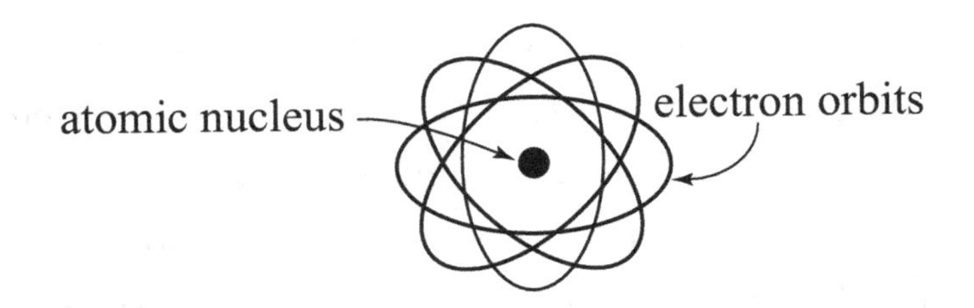

Figure 11.1: Model of an atom according to Ernest Rutherford. The drawing is not to scale because if the electron orbits were as large as shown in the drawing, the nucleus would be too small to see.

called α, β, and γ rays are emitted from the nuclei of certain atoms.

At first, physicists were reluctant to accept Rutherford's model because if electrons are to orbit the nucleus, they must accelerate (recall that a change in direction is acceleration). But acccording to Maxwell's theory, an accelerating charged particle emits electromagnetic radiation. The radiation contains energy, so the accelerating electrons should continually lose energy and spiral into the nucleus. But they do not. A way out of this impasse was given in 1913 by Niels Bohr (1885–1962), a Danish physicist, as we shall discuss in the last section of the next chapter.

There are, of course, many different kinds of atoms, like hydrogen, helium, carbon, and oxygen, to name a few. These atoms differ in that they have different values of electric charges in their nuclei. It was subsequently discovered that an atomic nucleus in general is composed of particles called protons and neutrons. A proton contains a positive electric charge equal in magnitude (and opposite in sign) to the negative electric charge on the electron. A neutron is electrically neutral. A proton or a neutron is almost 2000 times as massive as an electron, and that is why nearly all the mass of an atom is in its nucleus.

As an example, an ordinary carbon atom contains six protons and six neutrons in its nucleus, and is surrounded by six electrons, making the carbon atom as a whole electrically neutral. If a carbon atom loses an electron, as can easily happen, then it becomes a positively charged carbon "ion." If a carbon atom captures an additional electron, it becomes a negatively charged carbon ion.

In a nucleus composed of protons and neutrons, an additional force must be present besides the electromagnetic and gravitational forces. The reason is that the electromagnetic force among the positively-charged protons in a nucleus is repulsive and therefore tends to push the protons apart. The protons could not be bound together in a nucleus by a repulsive force, and the attractive gravitational force is far weaker than the repulsive electrical force. The neutrons, being electrically neutral, also would not be bound in nuclei by electromagnetic forces. Therefore, there must be an additional attractive force among the protons and neutrons, stronger than the electromagnetic force, to hold the nucleus together. This force is called the "strong force" or "strong interaction."

We earlier remarked that some atoms are unstable, usually emitting α, β, or γ rays (or particles). We also noted that β rays are electrons and γ rays are electromagnetic radiation. The doubly-charged α particles were determined to be the nuclei of ordinary helium atoms, each containing two protons and two neutrons and having the symbol ${}^4_2\text{He}$. All three kinds of radiation are now known to be emitted by the nuclei of various radioactive atoms.

The strong force is not only responsible for holding atomic nuclei together but in some cases permits unstable nuclei to emit α particles. It is the electromagnetic force that permits some unstable nuclei to emit γ rays. But it was determined that the emission of β rays requires the existence of a new force, the so-called "weak force" or "weak interaction." We shall discuss the strong and weak forces further in later chapters.

Chapter 12

Early Days of Quantum Theory

> It is wrong to think that the task of physics is to find out how nature is. Physics concerns what we can say about nature.
>
> —Niels Bohr (1885–1962)

In the late 19th century, the edifice of classical Newtonian physics, although on the whole remarkably successful, was confronted by several unexplained facts. These facts, which seemed at first to be no more than peripheral puzzles, led in the early 20th century to the overthrow of the classical picture. One of these revolutions, as we have seen, came with the theory of relativity, which overthrew the Newtonian idea that space and time had properties independent of the observer. The other revolution, which took longer, was the theory of quantum mechanics, which overthrew the notion that the position and momentum of a particle could be measured with arbitrary precision. From our quantum perspective, we regard the relativistic physics of Einstein as classical, in that it is not quantum physics, but Einsteinian physics is not Newtonian classical physics. Newtonian physics is often called nonrelativistic classical physics.

According to quantum mechanics, what we call particles have some of the properties of waves, and what we call waves have some of the properties of particles. Despite the overthrow of classical physics, it continues to be very successful in its domain of applicability. It is just that its domain is smaller than previously believed.

12.1 Blackbody radiation

If you put a steel poker into a fire, it will heat up. If you remove the poker and put your hand near it, you will feel heat radiating from the poker even if the poker is not hot enough to glow. The radiation is primarily infrared radiation, which can be felt but not seen by the human eye. If you hold the poker in the fire for a longer time, it will begin to glow, at first a dull red and then brighter. If the fire is hot enough, the poker will glow white hot. The color of the poker and the total electromagnetic radiation emitted by it depend on the temperature.

Now instead of a poker, consider a hollow steel ball with a tiny hole in it. The interior of the ball is black, as you can tell by putting your eye to the hole and looking inside. The inside of the ball is very close to what physicists call a "blackbody." An object at room temperature appears black if it does not reflect any of the light incident on it, but rather absorbs all the light. (If white light is incident on a red object, it reflects most of the red light and absorbs most of the other colors.) A truly blackbody is an idealization, but some objects come pretty close.

If the inside of our steel ball is hot, radiation will emerge from the hole. A graph of the intensity of the radiation, plotted against its frequency, is called the spectrum of the radiation. Because the frequency f is related to the wavelength λ by the formula $f\lambda = c$, where c is the speed of light, the intensity plotted against the wavelength is also called the spectrum. In Figure 12.1 we show a schematic drawing of the spectrum of wavelengths of light emitted by a blackbody at some arbitrary temperature. If the temperature is increased, the total intensity of the radiation will increase and the peak intensity will shift to a shorter wavelength. Conversely, if the temperature is decreased, the total intensity will decrease and the peak intensity will shift to a longer wavelength.

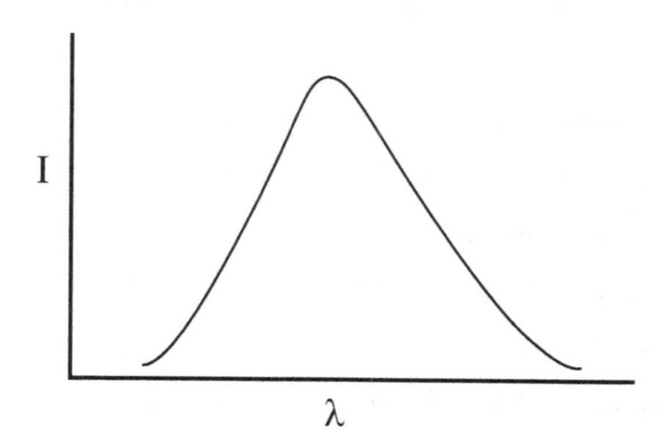

Figure 12.1: Spectrum of blackbody radiation, showing the intensity I plotted against the wavelength λ. If the blackbody is at a higher temperature, the peak intensity will increase and shift to the left.

The spectrum is called a "continuous" spectrum because the radiation is emitted over a whole range of wavelengths with no wavelengths in this range omitted. Using classical physics, physicists were able to show that the energy of the radiation emitted by a blackbody per unit area depends only on the temperature. The sun behaves as a blackbody with a surface temperature of more than 6000 kelvins (6000 K), a value similar to 6000 degrees Celsius. (Recall that 0 K is -273° C, which is only a small fraction of 6000.)

Physicists tried to calculate the value of the energy for any given temperature of the blackbody. The way physicists attempted to calculate the energy was by assuming a blackbody has a lot of tiny charged particles oscillating inside it, and these particles emit radiation of the same frequencies as the oscillation frequencies of the particles. The trouble was that when physicists tried to calculate the spectrum using classical physics, they obtained the wrong answer at high frequencies and the absurd answer that the total energy emitted was infinite.

In the last year of the 19th century, 1900, Max Planck (1858–1947), a German physicist, successfully calculated the spectrum of radiation from a blackbody. He did so by making what seemed like an arbitrary assumption: namely, that the tiny oscillators do not have arbitrary energies but only certain energies E_f, given by $E_f = hf$, where f is the frequency of an oscillator and h is a fundamental constant of nature. called Planck's constant. It has the dimension of an energy multiplied by a time and has the very small value $h = 6.63 \times 10^{-34}$ joules-seconds (J s). The number 10^{-34} means 1 divided by a number with 1 followed by 34 zeros. Planck's assumption seemed absurd to scientists at the time, but it was taken seriously because the theoretical calculation agreed with experiment. The reason that it took so long for Planck's constant to be discovered is that it is so small that its effects show up principally in microscopic phenomena.

Thus was born the quantum theory: a theory that says (among other things) that certain objects do not have arbitrary values of some quantities (such as energy) that can vary continuously but instead have certain special (quantized) values.

12.2 Photoelectric effect

As we have seen, it was shown that light exhibits many of the properties of a wave. Light of a certain frequency has a definite wavelength, and light exhibits properties of waves in undergoing interference and other wavelike phenomena.

Yet in the 19th century it was observed that light had a property that could not be explained in terms of the wave theory: the photoelectric effect. This effect is the phenomenon that when light of sufficiently short wavelength (or, in other words, of sufficiently high frequency) strikes a metal, electrons are knocked out of the metal. Of course, because light waves have energy, it could be assumed that the energy is sufficient to knock the electrons out of the metal. The problem was in the details.

If light is a wave, then an intense beam of light of a definite wavelength, having more energy than a faint beam of the same wavelength, ought to knock out electrons with a higher maximum energy than the faint beam. However, it was measured that the maximum energy of the emitted electrons was the same in both cases. A higher intensity of light just caused more electrons to be knocked out, but they did not have higher energy. It was also measured that the maximum energy of the electrons depends on the color (the wavelength) of the light—red light does not knock out any electrons at all, for example, while blue light does.

In 1905 Einstein explained these effects by assuming that, in the photoelectric effect, light acts like a bunch of particles rather than as a wave. These particles are called quanta of light, and were subsequently named "photons." Einstein assumed that each particle of light of frequency f has en energy E, which is related to its frequency by the formula $E = hf$, where h is Planck's constant. Einstein further assumed that an electron is knocked out of a metal by a single photon, which disappears in the process. The photon disappears because it is absorbed by the electron, which, by the conservation of energy, takes on all the energy of the photon. Since each photon has an energy that is proportional to its frequency (inversely proportional to its wavelength), the maximum energy of an emitted electron depends only on the frequency (or the wavelength) of the light, and not on its intensity. A photon of red light does not have sufficient energy to knock out any electrons, while a photon of blue light does.

Einstein in fact wrote down a formula that quantitatively explained the maximum energy of the emitted electrons. The formula says that the minimum energy required to knock an electron out of a metal (which is different for different metals) plus the maximum kinetic energy of the emitted electron is equal to the energy of the photon. It is a simple formula, making use of the principle of conservation of energy, and it agrees with the experiments.

It was a revolutionary idea that light could behave as particles, and it was many years before it was generally accepted by physicists. Even Planck, who first proposed the idea of a quantum of energy, did not at first believe Einstein's hypothesis.

A particle of light, or photon, has momentum as well as energy. As we remarked in Chapter 10 on relativity, the energy of any particle traveling at the speed of light is proportional to its momentum p, or $E = pc$, so this formula holds for photons. Because light is also a wave, even a single photon can be considered to have a wavelength. The wavelength λ of a photon is related to its momentum by the formula $\lambda = h/p$. This formula connects a wave property (wavelength) and a particle property (momentum) of a single photon.

12.3 Line spectra

When atoms are given energy, either as a result of collisions with other atoms or by absorbing light, the atoms soon radiate this energy away by emitting light. The big puzzle in the late 19th and early 20th centuries was that the radiation did not have a continuous spectrum, or, in other words, the radiation did not consist of light at a whole range of frequencies. Instead, the radiated light is emitted in one or more lines of definite frequency (and wavelength). This behavior of the radiation is completely unlike the radiation from a blackbody, which has a continuous spectrum only depending on the temperature. In Figure 12.2 we show part of the line spectrum of hydrogen. The intensities are high at certain definite wavelengths while no light is emitted at other wavelengths. We do not show the relative intensities of the different lines, because these intensities depend on the conditions.

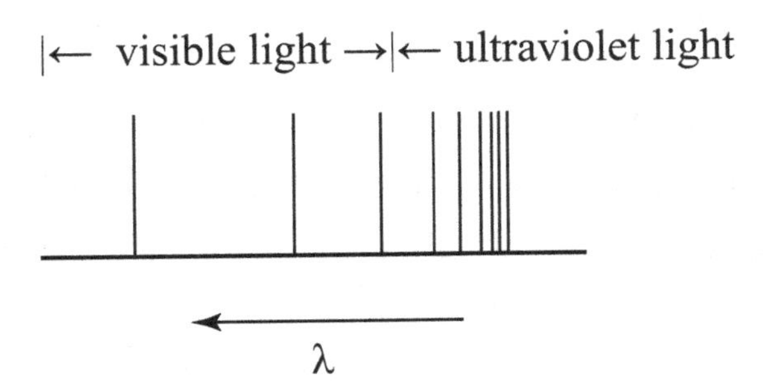

Figure 12.2: Part of the line spectrum of hydrogen. We show wavelengths at which light is emitted as lines, but we do not show the intensities of different lines, as these intensities depend on the conditions. Note that wavelengths increase to the left.

12.4 Bohr model of the atom

The first glimmer of understanding of line spectra came from the ideas of Niels Bohr (1885–1962), a Danish physicist. Bohr was one of the physicists who accepted Rutherford's result that an atom consists of a central nucleus surrounded by electrons in orbit around it. Bohr got around the problem of electrons radiating away their energy by suggesting that Maxwell's electromagnetic theory simply did not apply to the radiation from atoms. Instead, Bohr postulated that electrons in orbit around atomic nuclei could exist only in certain special orbits, each orbit having its own energy. Electrons, he said, only radiated when they made a transition from one orbit to another orbit with a different energy. If the energy of an electron in an initial orbit is E_i and its energy in the final orbit is E_f, then the energy of an emitted photon is simply the difference in energy between the intial and final orbits. Bohr assumed that the energy of the photon is related to its frequency by the relation $E = hf$, where again h is Planck's constant. Thus, the formula for the frequency of the emitted radiation is simply $hf = E_i - E_f$.

Bohr was not able to calculate the different possible energies of electrons in atoms containing more than one electron. However, in the case of the simplest kind of atom, hydrogen, which has only one electron going around its nucleus, Bohr was able to calculate the possible energy levels with the help of an additional assumption. This assumption was that the angular momentum L of an electron in an orbit could only have discrete, quantized values given by $L = n\hbar$, where the symbol $\hbar$ stands for Planck's constant h divided by 2π and n is an integer. The results of Bohr's calculations agreed very well with experiments. Again, quantum effects were showing up in the microscopic domain.

Chapter 13

Quantum Mechanics

> Two seemingly incompatible conceptions can each represent an aspect of the truth.
>
> —Louis de Broglie (1892–1987)

13.1 Particle waves

In 1924, Louis de Broglie (1892–1987), a French physicist, had the inspired idea that if light could behave as a particle as well as a wave, perhaps an electron could behave as a wave as well as a particle. We saw in the last chapter that the relation between the wave and particle nature of light is given by $\lambda = h/p$, where λ is the wavelength of the wave, p is the momentum of the particle, and h is Planck's constant. De Broglie postulated that the same relation holds for a particle, such as an electron. His inspired guess turned out to be right. It turns out that all kinds of objects that we call particles also have wave properties. There is a kind of wave-particle duality of microscopic objects.

Our intuition, as human beings, does not seem capable of understanding something that has properties of both particles and waves. Rather, our intuition can understand something that is *either* a particle or a wave, but not both. It is too bad for our intuition that although we observe things on the macroscopic scale that behave either like particles *or* like waves, but not both, we find objects on the microscopic scale that sometimes behave as particles and sometimes like waves. However, when we do an experiment that brings out the particle nature of a microscopic object, the experiment hides the wave nature, and vice versa. Bohr said that we cannot do an experiment that simultaneously shows the wave and particle nature of an object. His statement, called "the principle of complementarity," just seems to generalize what we have so far observed.

13.2 Nonrelativistic quantum mechanics

Not too long after de Broglie's work, Werner Heisenberg (1901–1976), German, and Erwin Schrödinger (1887–1961), Austrian, independently proposed a new theory, now called quantum mechanics, which accounted for both the wave and particle properties of matter. The formulations of the theory, as given by Heisenberg and Schrödinger, were different. Heisenberg's formulation involved mathematical quantities called "matrices," and is sometimes called "matrix mechanics." On the other hand, Schrödinger's formulation involved a "wave equation" for particles, and is sometimes called "wave mechanics." Later, Schrödinger proved that the seemingly different theories were equivalent, and now we usually just call the theory "quantum mechanics."

Both the Heisenberg and Schrödinger formulations apply to the case of nonrelativistic motion, that is, motion in which particles move much slower than the speed of light. Schrödinger also wrote down a relativistic wave equation, which satisfies the principles of special, but not general, relativity. Heisenberg's formulation can also be made compatible with special relativity. We defer discussion of relativistic quantum mechanics until later in this chapter.

We here discuss the Schrödinger formulation. He proposed a *wave equation* that would govern the motion of *particles*. The solution of Schrödinger's equation gives the value of a quantity called a "wave function," which has a different value at every point in space and time.

The Schrödinger equation has the peculiar property that it does not consist entirely of real quantities. Instead, the square root of minus one ($\sqrt{-1}$), which is called an imaginary number and given the symbol i, appears in the equation. (The square of a real number is always positive or zero, while the square of an imaginary number is negative.) A solution of Schrödinger's equation is not necessarily a real numbers but can be a "complex" number. A complex number contains both a real and an imaginary part; the imaginary part is multiplied by $\sqrt{-1}$. However, we can form a real number from a complex number by taking its "absolute value." The square of the absolute value is the square of the real part plus the square of the imaginary part (not including the $\sqrt{-1}$).

The interpretation of the wave function was given by Max Born (1882–1970), a German physicist. We first simplify by discussing the wave function associated with a single particle. In this case Born's idea was that the absolute value squared of the wave function at any point in space at any time represents the *probability* that the particle is at that point at that time. In the case that Schrödinger's equation describes more than one particle, the value of the wave function depends on the position of each particle. Then Born's interpretation says that the probability of each particle being at a certain place (different for each particle) is equal to the absolute value squared of the wave function when the various places are specified.

There are an infinite number of points in space, so the probability of finding a particle at a single point must be zero. However, from the wave function we can calculate the probability of finding a particle in a small region containing the point. This probability is the product of the size of the region and what is called the "probability density" at the point in space and time. The total probability of finding the particle somewhere at any time is the probability in each small region summed over all space. This total probaility must be unity (which means 100 percent), as the particle must be somewhere. There is an exception to this statement, as some particles can decay into other particles or can be annihilated. We discuss this exception in the section on relativistic quantum mechanics.

Barring the exception, the probability density in a small region in space can change with time if probability flows into or out of the region as time goes on. We call this flow of probability a "probability current." If there is no probability current, the probability in any small region remains constant as time goes on. If this is the case, we call the state of the particle a "stationary state."

In quantum mechanics, Bohr's allowed orbits for electrons in an atom are replaced by almost stationary states. The wave function of such a state is calculated from the Schrödinger equation. Except for the state of lowest energy (called the ground state), which usually is really stationary, an almost stationary state can radiate a photon and make a transition to another almost stationary state with lower energy or to the ground state. By solving the Schrödinger equation, physicists are able to calculate the frequency of light emitted by atoms containing more than one electron. However, as the number of electrons increases, the calculation becomes increasingly difficult, and approximation methods are necessary to solve the equation.

Schrödinger's equation describes the wave properties of matter, and, with Born's interpretation, the solution of the equation gives a particle interpretation of the wave function in terms of a probability. As a result, in general even if we know where a particle is at some initial time, we do not have the ability to make an exact prediction of where the particle will be at any future time. We seemingly had that ability in Newton's nonrelativistic classical mechanics, because, from Newton's equations of motion, given the position and velocity of a particle at a given time, we could calculate the position and velocity at any future time. However, Newton's laws of motion are only approximately correct. Even at low particle speed, where we do not need to take relativity into account, Newton's laws do not account properly for the motion of microscopic particles. Under many circumstances, Newton's laws provide an excellent description of the motion of macroscopic systems, but we must not forget that the description is approximate and can in certain circumstances break down completely. In the next section we describe a reason for the breakdown.

13.3 Heisenberg's uncertainty relations

It was not too long after the invention of quantum mechanics that Heisenberg showed that so-called *uncertainty relations* could be deduced from the theory. We shall discuss some of the uncertainty relations for a single particle. Suppose the position of the particle is somewhat near the position specified by the coordinate x, at a time close to t. Furthermore, suppose the particle has a momentum in the x direction not too far from p and that it has energy near E. Let us call "the error (or uncertainty) in a measurement of" by the symbol Δ. Then Δx is the error in a measurement of the position along the x axis, Δp the error in a measurement of the momentum p, and so on. Heisenberg's uncertainty relations for these quantities are $\Delta x \Delta p \geq \hbar/2$ and $\Delta t \Delta E \geq \hbar/2$. Here the symbol $\geq$ means "greater than or equal to" and $\hbar$ is Planck's constant h divided by 2π. There are analogous uncertainty relations for the positions and momenta along the y and z axes and for certain other pairs of quantities as well.

What Heisenberg's uncertainty relations tell us is that no matter how good our measuring aparatus is, it is impossible to measure, for example, both the position and the momentum of a particle to arbitrary accuracy. The product of the uncertainty in the measurements of the position and the momentum must be at least as great as Planck's constant divided by 4π. The better the measurement of the position, the worse will be the measurement of the momentum, and vice versa. Likewise, the more precisely we measure the time at which we observe a particle, the poorer will be our measurement of the energy of the particle and vice versa.

In Newtonian mechanics it was *assumed* that we could measure both the position and the velocity of a particle with arbitrary accuracy. We see that the theory of quantum mechanics says it isn't so. Why, then, did the Newtonian assumption go unchallenged for more than 200 years? The answer is that Planck's constant is very small and the mass of a macroscopic particle is very large (compared to the mass of an electron). The mass of the particle is relevant because momentum is equal to mass times velocity ($p = mv$). Suppose the position of a particle is measured with only a small error. Then, by Heisenberg's uncertainty principle, the error in the momentum must be large. But if the mass of the particle is large, then, because of the equation $p = mv$, even a small error in the velocity leads to a large error in the momentum because the small error in the velocity is multiplied by a large quantity: the mass. For a large enough mass, the minimum error in the velocity will usually be too small to observe, so that Newtonian mechanics becomes an excellent approximation to quantum mechanics.

The situation is different for a particle of small mass, like the electron. The mass of the electron m_e has been measured to be $m_e = 9.11 \times 10^{-31}$ kg. If an electron is in an atom and we try to measure its position to a much smaller region of space than the atom itself, the uncertainty in its momentum becomes so large that the electron probably escapes from the atom.

Of course, even in classical mechanics, we do not measure position and momentum to infinite accuracy. The point is that the theory of classical mechanics places no *intrinsic* limitation on the accuracy with which we can make measurements. If classical mechanics were correct, then we would be able to improve our measuring apparatus without any theoretical limitation. However, quantum mechanics says that, no matter how ingenious we are, there is no way we can build a measuring apparatus good enough to violate the Heisenberg uncertainty relations.

Let us consider a measurement of position in some detail in order to get a rough idea of how the uncertainty relations arise. In order for us to measure the position of a small particle, we must look at it. What this means is that we shine light on the particle and let the light bounce back into our eyes. But light consists of photons, which have momentum, and the photons impart some of their momentum to the observed particle, leading to an error in a measurement of the momentum of the particle. If we use light, or any wave, we cannot locate the postition of a particle to a precision much smaller than the wavelength of the light we use. Therefore, the more precisely we want to measure the position, the shorter the wavelength of the light must be, which implies the larger the momentum of each photon must be in view of the relation $p = h/\lambda$. And the larger the photon's momentum, the greater the momentum it can impart to the measured particle, leading to a larger error in the particle's momentum. Careful analysis leads to the conclusion that if quantum mechanics is correct, then Heisenberg's uncertainty relations cannot be violated.

13.4 Interference in quantum mechanics

If we shine light through a thin slit in an opaque object and place a screen behind the slit, then the light will form a pattern on the screen. Because light is a wave, it will bend as it goes through the slit (we call this phenomenon "diffraction"), and the image on the screen will be wider than the slit itself. We can also shine light through a double slit, and again we get a pattern on the screen. However, the pattern is not simply the sum of the patterns caused by the individual slits. Because light is a wave, the light going through the two slits exhibits interference. In particular, there are regions on the screen that are illuminated if either slit is closed, but the regions are dark when both slits are open. the reason is that the wave exhibits destructive interference in those regions.

Let us now make the light source so dim that only one photon goes through the slits at a time and makes a tiny spot on the screen as if it is a particle. Suppose the spots on the screen are recorded as more and more photons impinge on it. The pattern of recordings is the same as the diffraction pattern from two slits. Those who say that each photon must go through only one slit get the wrong answer, because they predict, in contrast to experiment, that the pattern is the sum of the patterns from one slit at a time.

The only way we can explain the phenomenon is to say that even a single photon acts as a wave when it goes through the slits, somehow "being aware" of both slits as it passes through, but the photon acts as a particle when it makes a spot on the screen.

If we do the same experiment with a beam of electrons going

through two slits and hitting a screen that scintillates when an electron strikes, we get the same answer as with a beam of photons. The conclusion is that an electron acts as a wave when it goes through the slits and it acts like a particle when it hits the screen.

The question, "Which slit did the photon or electron go through?" cannot be answered. According to quantum mechanics, the question does not make sense unless an experiment is performed to measure which slit the photon or electron goes through. It is a difficult measurement to perform in practice, but it can be analysed in principle. Such an experiment is called a "thought experiment." The result of the analysis is that if a measurment determines which slit the photon or electron goes through, the interference phenomenon is destroyed, and the pattern on the screen is the result of the sum of the patterns with a single slit open at a time. In Figure 13.1 we illustrate the diffraction pattern after a wave (of light or electons) passes through a single slit. We also illustrate the interference pattern after the wave passes through two slits.

If a two-slit interference pattern is observed, one cannot say that the photons or electrons "really" go through either one slit or the other. They act as waves that pass through both slits at the same time. Our intuition does not allow us to "understand" how photons and electrons can act both as waves and particles. There is no inherent contradiction in the theory of quantum mechanics, as far as we know, and furthermore, the calculations of quantum mechanics lead to predictions that agree with experimental measurements. But the human mind does not seem to be able to comprehend what is "really" going on.

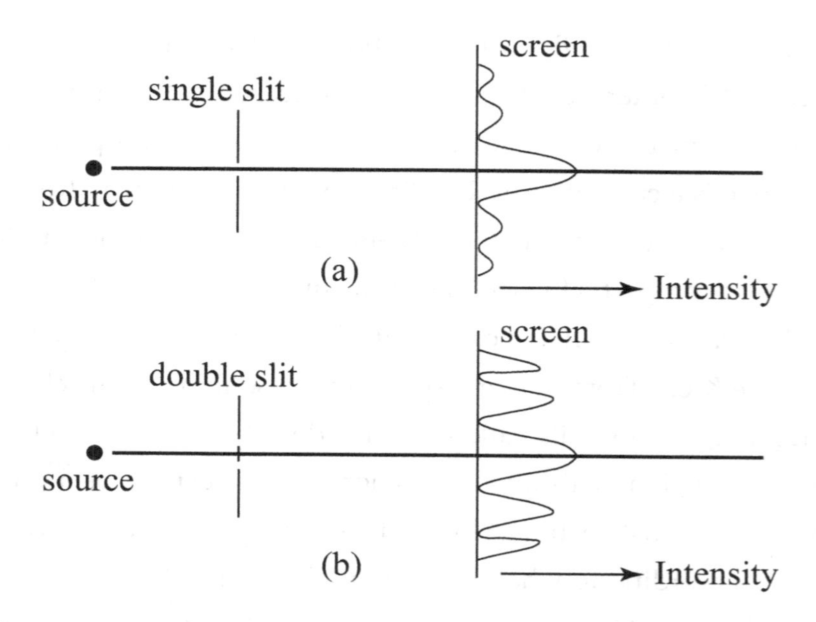

Figure 13.1: Diffraction and interference. (a) Pattern made by light or electrons going through a single slit. (b) Pattern made by light or elections going through a double slit.

13.5 Relativistic quantum mechanics

At about the same time as Schrödinger wrote down his successful nonrelativistic equation, he wrote down another wave equation compatible with the principle of special relativity. This equation is occasionally referred to as Schrödinger's relativistic equation but, for historical reasons that I do not appreciate, is usually referred to as the Klein–Gordon equation. Schrödinger's nonrelativistic equation is usually just called the Schrödinger equation.

Recall that Born interpreted the absolute square of the wave function solution to the Schrödinger equation as a probability. However, the absolute square of the wave function of the Klein–Gordon equation cannot be interpreted that way. The reason is that a probability density must have the property that it changes at a point only when there is a current of probability moving into or away from the point. The absolute square of the Klein–Gordon wave function does not have the property. In fact, there is no quantity that can be formed from the wave function that can be interpreted as a probability. However, if the Klein–Gordon is interpreted as describing a particle with electric charge, a quantity called a charge density can be formed from the wave function. This quantity is not always positive, as a probability density must be, but a charge density may be either positive or negative, according to whether the particle has positive or negative electric charge. With this interpretation, the Klein–Gordon equation describes a charged, but not a neutral, particle. However, a different interpretation may be given, in which the Klein–Gordon equation is supposed to describe many particles. We do not describe this interpretation in detail, but shall say some words about it after we discuss a different relativistic wave equation: the Dirac equation.

In 1928 Paul Dirac (1902–1984), an English physicist, wrote down a relativistic wave equation that differed from the Klein–Gordon equation. A solution to the Dirac equation turns out to provide a good description of the electron.

According to the solution of the Dirac equation for an electron, an electron, whether in motion or at rest, has an intrinsic angular momentum, called "spin" and denoted by the symbol S. It may be helpful to think of the electron spinning on its axis like a top, but that is a poor analogy. Nobody can really picture the spin of the electron because, to the best of our knowledge, an electron behaves like a particle that is no bigger than a point. It is hard to visualize a point spinning. We think the electron might be a point because all attempts so far to measure the size of an electron have failed. The measurements have only been able to determine that an electron is smaller than the smallest size we can measure. Of course, this does not necessarily mean that no future measurement will be able to measure that an electron has a size bigger than zero. We shall have to wait and see. Furthermore, some theories, such as so-called "string" theories, give an electron a size bigger than zero. However, string theories so far have not been confirmed by experiment, and so at the present time must be regarded as speculative. We briefly discuss string theories in the final chapter.

All electrons behave as completely identical particles, and therefore, all have the same magnitude of the spin. Only the direction of the spin can differ for different electtrons, and even the direction is constrained by the Dirac equation. If the electron is spinning clockwise around some direction in space, then we say that its spin is pointing along the direction. If the electron is spinning counterclockwise, we say its spin is pointing opposite to the direction we have chosen. It turns out that a precise measurement of the component of the spin of any electron along any arbitrary direction in space can give only one of two values, either along the direction or in the opposite direction, with the magnitude $S = \hbar/2$. This prediction of the Dirac equation has been confirmed by experiment. We use a shorthand notation by saying that all electrons have spin $1/2$ in units of $\hbar$, which is the same as omitting the $\hbar$.

The Dirac equation has other astonishing properties. For example, it not only has solutions corresponding to an electron with a positive energy but also solutions in which the electron has a negative energy. We have seen that an electron in an excited state in an atom can spontaneously make a transition to a lower-energy state. Why, then, cannot a free Dirac electron spontaneously make a transition to a negative energy state? Dirac's solution was to say that all the negative-energy states are filled with a "sea" of electrons.

It turns out that a positive-energy electron cannot make a transition to a state already occupied by a negative-energy electron. The fact that no two electrons may occupy the same quantum state is called the "Pauli exclusion principle" or just the "Pauli principle" after the Austrian physicist Wolfgang Pauli (1900–1958), who first proposed it. We discuss the Pauli principle in more detail later in this chapter.

But why do we not observe all the negative-energy electrons in the sea? Dirac audaciously assumed that we cannot not observe the negative-energy electrons in the sea, but interpret the filled sea as a vacuum. However, if a sufficiently energetic photon is absorbed by a negative-energy electron in the sea, the electron can be excited to a positive-energy state, leaving a "hole" in the sea. Dirac interpreted the hole as as an "antiparticle" with positive energy and opposite electric charge to the particle. Because the electron has negative charge, its antiparticle must have positive charge. At first, Dirac hoped that the antiparticle was the proton, but soon it became clear from the theory that the antiparticle must have the same mass as the particle. The antielectron is usually called a "positron." The prediction of the existence of the positron was made before it was discovered in 1932, with positive charge and the same mass as the electron. This was a triumph for Dirac's theory.

It turns out that the property of having spin and having an antiparticle is not unique to the electron. All particles that we presently (and tentatively) call elementary are known to have definite values of spin and to have antiparticles, although the spin of a particular particle may be zero, and the antiparticle of an electrically neutral particle may be the same as the particle.

Because Dirac's equation seemingly required a filled sea of negative-energy electrons in order to be consistent, the equation could no longer be interpreted as merely an equation for a single particle. A later interpretation of Dirac's equation that does away with the requirement of the sea is that the equation does not simply describe a single particle but is the equation of a field. The Dirac field can be quantized, but the procedure for doing so is highly technical, and we cannot go into it. When the field is quantized, electrons and positrons are the quanta, so that in the field interpretation, the equation in general describes more than one particle.

Maxwell's equations are compatible with the special theory of relativity. Among the solutions of Maxwell's equations is a description of the electromagnetic field. This field describes light as an electromagnetic wave. We can go beyond Maxwell's equations and quantize the field. The procedure for quantizing the electromagnetic field is also highly technical, but the result is that we can describe the electromagnetic feld in terms of photons. If we do this, we find that photons, like electrons, have spin, except that it turns out from the equations that photons have spin 1, again in units of $\hbar$). Photons are electrically neutral (uncharged), and have no properties that would distinguish a particle from an antiparticle. Therefore, photons are identical to their antiparticles.

With the quantization of Maxwell's equations for the electromagnetic field and the quantization of the Dirac field, we find that the equations of relativistic quantum mechanics are equations for many particles, photons and electrons (and positrons) rather than each being an equation for a single particle.

We can go further, and write down an equation that describes the interaction of the quantized Dirac and electromagnetic fields. The solutions show that electrons (and positrons), being charged particles, can either emit or absorb photons one at a time. Furthermore, a photon (in the neighborhood of an atom or other particle) can turn into an electron and a positron, and an electron and positron can collide, annhiliating into two photons. All these predictions of the theory have been borne out by careful experiments. Thus, according to relativistic quantum mechanics, the number of photons is not a constant, as photons can be created and absorbed by electrons (and by any other charged particle). Also, while an electron or positron cannot singly appear or disappear as a consequence of its interactions with photons, an electron-positron pair can either be created or annihilated. This leads us to a new conservation law: the number of electrons minus the number of positrons is conserved, i.e., remains constant as time goes on. This law is only approximate, as an interaction that we have discussed only briefly, the so-called "weak" interaction, can lead to the creation or disappearance of a single electron, although electric charge is conserved in weak interactions.

As an example of a weak interaction, consider a free neutron. One may obtain a free neutron by using an energetic particle to knock a neutron out of an atomic nucleus. It has been observed that free neutrons are unstable, decaying by the following reaction:

$$n \to p + e^- + \bar{\nu},$$

where n stands for a neutron, p for a proton, e^- for an electron, and ν for a neutrino. The bar on the symbol for a particle stands for its antiparticle, so that $\bar{\nu}$ stands for an antineutrino. The neutrino and antineutrino are electrically neutral, and have only weak interactions, as far as we know. In the decay of a neutron, the emitted antineutrino has such weak interactions that it is normally not observed. Electric charge is conserved because the initial neutron has charge zero, and the charges of the final proton and electron add algebraically to zero.

Before the discovery of the neutron, it was observed that some heavier kinds of nuclei decay into an electron and a residual nucleus. The residual nucleus has a greater positive charge than the original decaying nucleus, because the emitted electron has negative charge and the total charge is conserved in the reaction. The decay of a radioactive nucleus results in the transmutation of one element into another. Such transmutation was the dream of the alchemists, although in radioactive decay, the transmutation is uncontrolled and usually involves only tiny amounts of the decaying element.

In the early observations of the decay of a nucleus into an electron and a residual nucleus, it was noted that the sum of the energies (including rest energies) of the residual nucleus and the electron was less than the rest energy of the decaying nucleus. At first, this fact caused much concern, and it was suggested by prominent scientists that energy was not conserved in the decays. Wolfgang Pauli came up with a different hypothesis: that a neutral, unseen particle, was also emitted in the decay, and carried away just enough energy to conserve energy overall.

It was the Italian physicist Enrico Fermi (1901–1954) who named Pauli's particle the neutrino. Fermi also proposed a theory of weak interactions that agreed with the early experiments, but his theory was superseded by the standard model. Fermi was unusual among 20th century physicists because he excelled in both experimental and theoretical work. Much more common in that century, and continuing into the 21st century, has been the specialization of physicists into either theory or experiment.

Because a neutrino is electrically neutral and because its interactions are weak, we still do not really know whether a neutrino is different from an antineutrino or whether a neutrino and an antineutrino are identical.

How long does a free neutron exist, or as we usually say, how long does a fee neutron "live"? The theory does not allow us to say. Some neutrons live longer than others, and a neutron may decay at any time. However, it has been measured that neutrons live 15 minutes *on the average*. This means that if we look at a large number of neutrons, we shall find that the average time for them to decay is 15 minutes, although we cannot say when any single neutron will decay. Thus we say that the neutron has a "mean life," or simply a lifetime, of 15 minutes. The mean life is a synonym for the average life.

We can also talk about a "half-life" of a decaying particle, or the time it takes for half of a collection of particles of a certain kind to decay. Because more particles decay at times shorter than the mean life than at longer times, the half-life of a particle is shorter than its mean life. The half-life of a neutron is about 10 minutes. The mean life is longer than the half-life because a few of a collection of decaying particles live much longer than the average, and this fact increases the average life. In fact, the mean life is about 44 per cent longer than the half-life.

There is a conservation law that, as far as we know, is exact, and that is the conservation of electric charge. Therefore, if an electron is created in a weak interaction, another particle with the same charge as the electron must disappear, or another particle with opposite charge of the electron must be created simultaneously. Likewise, if an electron disappears in a weak interaction, another particle with the same charge as the electron must be created, or an oppositely charged particle must also disappear. In the decay of the neutron, the positively charged proton and negatively charged electron are simultaneously created.

The theory that describes photons and electrons as a result of quantization of the electromagnetic field and the Dirac field is called "quantum electrodynamics." It is a very successful theory, enabling physicists to calculate some properties of electrons to better than one part in a million and agreeing with the results of experiments to this precision.

We next return to the Klein–Gordon equation. We saw that this equation does not describe a single neutral particle because a positive probability density cannot in general be constructed from the wave function. However, we can reinterpret the Klein–Gordon equation as the equation for a field that can be quantized. If the quantization is carried out, the resulting particles have spin zero. (That means the particles don't spin at all.) The particles may be either neutral or charged; if they are charged, then they interact with the electromagnetic field.

It is interesting that we have not been able to construct a field to describe protons or neutrons that is similar to the Dirac field for electrons. It turns out that the reason is that protons and neutrons are not elementary particles but are composites of still other particles called "quarks." We shall discuss the field for quarks in Chapter 16. Nevertheless, protons and neutrons have antiparticles, called antiprotons and antineutrons respectively. A proton-antiproton pair can be created or annihilated, and the same is true for a neutron-antineutron pair. We have observed these events in the laboratory. A proton and a neutron each have spin 1/2 (again in units of $\hbar$.)

We have learned from Maxwell's equations that if a charged particle moves, giving rise to a current, a magnetic field is created. Therefore, because electrons and protons are charged, their spins cause the charges to move, giving rise to magnetic fields. The magnitude of the magnetic field of the electron is predicted to an excellent approximation by the Dirac equation and to an even better approximation by the theory of quantum electrodynamics. The same cannot be said for the proton because it is a composite of quarks. Although the neutron is electrically neutral, its spin still gives rise to a magnetic field. The reason is that the neutron is a composite of quarks, which have both charge and spin, and so give rise to magnetism in the neutron. Because quarks are charged, their antiparticles (antiquarks) must have opposite charge and so must be different from quarks. Although a neutron is neutral, an antineutron is different from a neutron because an antineutron is composed of antiquarks and the direction of its magnetism is opposite to the magnetism of a neutron (relative to the direction of its spin).

13.6 Spin and statistics

According to quantum field theory, particles fall into two classes: those that have half-integral spin, commonly spin $1/2$ in units of $\hbar$, and those that have integral spin, commonly spin zero and spin one in units of $\hbar$. (Recall that the quantity $\hbar$ is defined to be equal to $h/(2\pi)$), where h is Planck's constant. The spin $1/2$ fields are called matter fields. The spin one fields are called force fields or interaction fields, and they carry the interaction between the matter fields. The spin zero field is a special field that is supposed to give mass to those quanta that have mass greater than zero. We say more about these fields later.

Consider several identical quanta of half-integral spin, such as electrons. Quantum field theory is only self consistent for such particles if the wave function of any two of them is antisymmetric under the operation of interchaning their coordinates. That means the wave function changes sign under the interchange. As a result, only one quantum can be in a given quantum state. We see this as follows: if two electrons are in the same state, interchanging them cannot change the wave function because all the coordinates describing the two electrons are the same. But the wave function must change sign under the interchange. But a wave function that must stay the same and must change sign can only be zero, which says the state cannot exist. Because no two identical particles of half-integral spin can exist in the same quantum state, the so-called "statistical" properties of a large collection of such particles are quite unusual. A collection of identical particles of half-integral spin, no two of which can be in the same quantum state, is said to obey Fermi–Dirac statistics (after the physicists Enrico Fermi and Paul Dirac). Particles of half-integral spin are called "fermions." The fact that no two electrons in an atom can be in the same quantum state is called the Pauli principle, as we have already mentioned. The properties of the chemical elements cannot be understood unless the Pauli principle is taken into account.

Consider several identical quanta of integral spin, such as photons. Quantum field theory is only consistent if the wave function for such particles is symmetric under the operation of interchanging their coordinates. That means the wave function does not change sign under the interchange. Thus, no contradiction occurs if two or more of the photons are in the same state. As a result, any number of identical integral-spin particles may be in the same state. A collection of such particles is said to obey Bose–Einstein statistics (after Einstein and the Indian physicist Satyendranath Bose). Particles of integral spin are called "bosons."

13.7 Entanglement

We consider a system of two electrons. As we have said, each electron has spin 1/2. The rules of quantum mechanics for combining angular momentum, which we shall not go into here, says that in this case the combined spin of the two electrons can be either 1 or 0, but nothing else. The combined spin is 1 if the spins of the two electrons are aligned in the same direction and 0 if the spins are oriented in opposite directions. We recall that if the direction of spin of a single electron is measured with respect to any axis in space, the result must be that the electron's spin is either along that direction or opposite that direction. If we don't have any knowledge beforehand of the direction of the spin, the probability that the spin is along the direction of measurement is 50 percent.

Now suppose the two electrons with total spin 0 are made to fly apart in opposite directions, and observers who are far apart measure the direction of each electron's spin. Suppose this experiment is repeated many times with different pairs of electrons. Each observer will measure the spin of the electron to be pointing along the direction of his arbitrarily chosen axis 50 percent of the time and pointing opposite to this direction 50 percent of the time. What we mean is that in the repeated experiments with the different electrons, 50 percent of the electrons will be measured to have their spins along the direction chosen and 50 percent will be measured to have the spins opposite the chosen direction.

Suppose by agreement each observer chooses the same axis along which to measure the electron's spin direction. Then although each has a 50 percent chance to measure the electron's spin as along or opposite the direction, if one observer measures the spin of an electron to be along his direction, he can instantaneously predict with certainty (100 percent probability) that the other observer will measure the spin to be opposite that direction. No matter how far apart the two observers are from each other, the prediction will be valid if nothing intervenes along the electrons' paths to change the direction of either of their spins.

What is happening is that a single wave function describes the spin of the two electrons no matter how far apart the electrons are in space. The spread-out nature of this wave function means that the direction of the spins of the electrons are not independent but *entangled* with each other.

At first, this 100 percent correlation between the observations of two distant observers of the spins of two different elections may not seem astonishing. Superficially, there is a classical analogue, but the analogue is not sufficient to explain what is going on in the quantum mechanical case. First we describe the classical analogue and then show how the quantum mechanical situation is different.

We consider a boy and girl sent off in opposite directions toward two observers along the x axis. The direction of the girl is chosen at random, but the boy always goes in the opposite direction. Then each observer will have a 50 percent probability of seeing a girl and 50 percent probability of seeing a boy. But if one observer sees a girl, he instantly knows with certainty that the other observer will see a boy.

The difference with quantum mechanics arises because an electron spin can be measured to be along or opposite *any arbitrarily* chosen axis. If one observer measures the spin to be along one axis, he can predict with certainty that the other observer will measure the spin to be opposite his chosen axis, provided the other observer chooses the same axis. However, if the other observer chooses a different axis, the first observer cannot predict with certainty whether the second observer will find the spin along or opposite his chosen axis. In particular, if the second observer chooses an axis that is perpendicular to the first axis, the first observer, despite his own measurement, will have only a 50 percent probability of predicting what the second observer will measure. In other words, the measurement of the first observer forces the spin of the first election to be either along or opposite any direction he chooses. That measurment also forces the spin of the second electron to be in the opposite direction to the first if the second observer chooses the same axis for measurement. But if the second observer chooses a perpendicular axis for measurement, the result of the first measurement puts no constraint at all on the result of the second measurment. The second observer will have a 50 percent chance of measuring the spin along his chosen axis and 50 percent chance of measuring the spin to be opposite that axis.

In the classical case of the boy and girl, which of the two observers will see the girl and which the boy is predetermined by the person who sent them off (or by the boy and girl themselves), even though each observer doesn't know in advance whether he will see the boy or girl. In the case of the electron spins, however, the person who prepares the two electrons in a state of total spin 0 doesn't know in advance whether the spin of an electron will be measured to be along or opposite any arbitrarily chosen axis. In fact, the electron spins cannot be said to be along or opposite any arbitary axis until *after* the first measurement. Before that measurement, the electrons have the potential to be along or opposite any axis at all.

It is a remarkable fact that a measurement of the direction of spin of one electron *immediately* lets one predict the direction of spin of the other election with respect to a common axis, no matter how far away the second election is from the first at the time of measurement. This fact is a prediction of quantum mechanics and has been confirmed by experiment. The immediate effect arises from a *nonlocality* of the wave function—the wave function is spread out over space. Einstein called this nonlocal effect, "Spooky action at a distance." The action may be spooky, but it doesn't violate special relativity. After the first observer makes a measurement, he knows what the second observer will find, but if he wants to tell the second observer, he will have to send a signal, which cannot go faster than the speed of light.

Chapter 14

The Elements

> There will come a time, when the world will be filled with one science, one truth, one industry, one brotherhood, one friendship with nature...
>
> —Dmitri Mendeleev (1834–1907)

14.1 Quantum restrictions on angular momentum

We have seen in the last chapter that the quanta of fields are particles with spin. The spin of a particle can be half-integral (usually $1/2$) or integral (usually 0 or 1). It follow from quantum mechanics that not only are the spins quantized (to half-integer or integer values) but also the orientation of the spins with respect to an arbitrarily chosen axis are also quantized. The value of the component of the spin along any axis can vary from the value of the spin down to minus the value of the spin, in unit steps (all in units of $\hbar$, of course). For example, the orientation of the spin of an electron, which has spin $1/2$, can be $+1/2$ or $-1/2$. The orientation of the spin of a particle of spin 1 can be $+1$, 0, or -1. A particle of spin 2 can have orientation $+2$, $+1$, 0, -1, -2. No matter what arbitrary axis is chosen as a measurement axis, these are the only allowed values of the orientation of the spin.

There is an exception to this rule. A particle with zero mass can have orientation either $+S$ or $-S$ with respect to an axis along its direction of motion, but not intermediate values. (Rememember, a particle with zero mass is always in motion at the speed of light.) For example, a photon, which has spin 1 and, as far as we know, is massless, can have orientation either $+1$ or -1 but not zero. If gravity can be quantized (this has not been accomplished if the classical theory is general relativity), then a massless quantum of the gravitational field, called a graviton, should exist. It should have spin 2, and its orientation with respect to an axis along its direction of motion should be either $+2$ or -2 but nothing else. Gravitons have not been detected, but in view of the weakness of the gravitational interaction, this is not surprising.

Not all particles are elementary. For example, an atom is a bound composite state of an atomic nucleus and one or more electrons. In a neutral atom, the number of negatively charged electrons equals the number of protons in the nucleus. What is the spin of an atom? The answer to this question depends on the dynamics, that is, on the forces between the particles (electrons and nucleons) that comprise the atom. But the answer also depends on the restrictions of quantum mechanics. Because these restrictions are quite general, independent of the dynamics, we shall discuss them in some detail.

Quantum mechanics says that if a particle with spin S is a composite of two particles of spin S_1 and S_2 and the particles do not have any orbital angular momentum, then the spin S is restricted to a set of values, the largest of which is $S_1 + S_2$, Other allowed values are $S_1 + S_2 - 1$ and down in unit steps to the absolute value of $S_1 - S_2$. For example, a hydrogen atom is a bound (composite) state of a proton with spin 1/2 and an electron with spin 1/2. Therefore, if the particles don't have orbital angular momentum, the spin of the hydrogen atom is either 1 or 0. The dynamics of the forces between the electron and proton make the lowest-energy state (called the ground state) have spin 0. The state of spin 1 can exist, but it has a higher energy.

Orbital angular momentum is also quantized. Its value can be an integer (in units of $\hbar$) but not half integral. If the electron in a hydrogen atom has orbital angular momentum (denoted by L), then the total angular momentum of the hydrogen atom is the vector "sum" of the orbital and spin angular momentum, "added" in the same way that spins are added. For example, if $L = 2$ and $S = 1$, then the total angular momentum of the hydrogen atom (denoted by J) can be 3, 2, or 1. If $S = 0$, then the total angular momentum is just the orbital angular momentum, which is 2 in our example. Because of the dynamics, the states with different total angular momentum J have different energies. We are here only stating results of the mathematics, as a detailed discussion of the mathematics is beyond the scope of this work.

14.2 Building up the elements

As we have remarked, an atom consists of a nucleus surrounded by electrons. Because an atom is electrically neutral (if it is not neutral, it is called an ion), the number of electrons in an atom is equal to the number of protons in its nucleus.

Substances composed of atoms of a single kind are called "elements," while substances that are formed from chemical binding of more than one element are called "compounds." The smallest amount of an element is a single atom. A molecule is the smallest amount of a substance (element or compound) that has the physical and chemical properties of the substance. The simplest molecule of some elements contains only one atom of the substance, but a molecule may be more complicated. For example, the smallest amount of ordinary oxygen gas is a molecule containing two atoms of oxygen. A single atom of oxygen has different properties from a molecule of oxygen, the atom being more reactive. An example of a molecule containing two different kinds of atom is sodium chloride, which contains an atom of sodium chemically bound to an atom of chlorine. A compound can have very different properties from the atoms of which it is formed. For example, sodium is a reactive metal and chlorine is a noxious gas, but sodium chloride is ordinary salt. We discuss chemical bonding in the next section.

The simplest atomic nucleus consists of one proton. To be electrically neutral, one elctron must be bound in the atom outside the nucleus. This atom is called hydrogen, and has the symbol ${}^{1}_{1}H$. The left superscript 1 stands for the number of nucleons (protons plus neutrons) in the nucleus, and the left subscript 1 stands for the number of protons in the nucleus. The letter H is the chemical symbol for hydrogen. The symbol for an atom just depends on the number of protons in its nucleus, and, because of this fact, the subscript is often omitted. Thus, the symbol for hydrogen may be written ^{1}H. Sometimes even the superscript is omitted, and the symbol for hydrogen is written simply as H.

If an atomic nucleus has one proton and one neutron, the atom is still hydrogen but is called either "heavy hydrogen" or "deuterium," with the symbol $^{2}_{1}$H, as opposed to ordinary hydrogen, which has no neutrons in its nucleus. Ordinary hydrogen and deuterium are called "isotopes" of the same element. Isotopes are defined to be atoms containing the same number of protons but different numbers of neutrons. There exists one other isotope of hydrogen with two neutrons in its nucleus, but it is unstable, which means it spontaneously decays into something else with a characteristic average life. The isotope $^{3}_{1}$H is called "tritium." In nature, most hydrogen is ordinary hydrogen.

The next simplest atom is helium, with two protons in its nucleus. The most common isotope of helium contains two neutrons in its nucleus, and has the symbol $^{4}_{2}$He. The nucleons move inside the nucleus, but are on the average close together. In Figure 14.1 we give a schematic drawing of a ^{4}He nucleus, showing that the individual protons and neutrons are near one another. The protons and neutrons are themselves composites of quarks, not shown. On this scale, the electrons are too far away to be drawn.

The lowest energy state of helium (the ground state) has two electrons around its nucleus, each electron being in its lowest energy state. But recall that the Pauli principle forbids both electrons from being in the same state. Therefore, one electron must have spin "up"

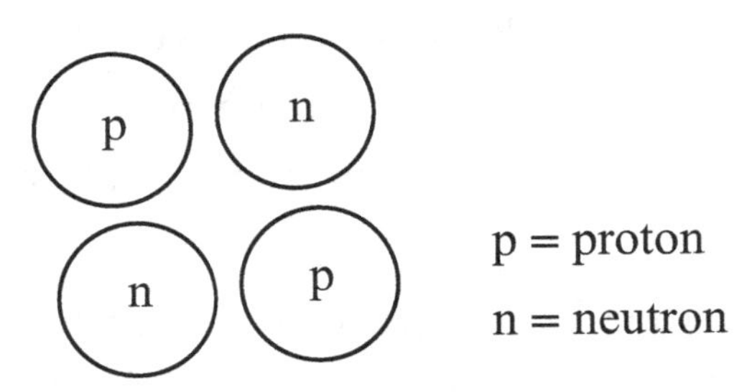

Figure 14.1: Schematic drawing of a $^4_2$He nucleus. This nucleus is also known as an alpha particle when it is emitted in radioactive decay. The protons are denoted by circles with p inside them, and the neutrons are denoted by circles with n inside them.

(along any arbitrarily chosen axis) while the other must have spin "down" (in the opposite direction along the chosen axis).

The atom with three protons in its nucleus is called lithium and its most common isotope is $^7_3$Li. The neutral atom has three electrons around it. Two of these electrons are in the same state as the electrons of helium, but the third electron, because of the Pauli principle, must be in a different state. The lowest energy state of the third electron has a different so-called "principal" quantum number from the other two, which means that its wave function depends on distance from the nucleus in a different way from the wave functions of the first two.

Because only two electrons can have the simplest wave function with low energy, and the third must have a higher energy, the first two electrons complete what is known as the "first shell" in the atom. The atom helium has what is known as a "closed shell," and this element is an inert gas. The atom lithium has one electron in a "second shell." The atom beryllium, which has four protons in its nucleus, has two electrons in its second shell.

Next comes the atom boron, which has five protons in its nucleus. One might think that the fifth electron would have to start a third shell, but that is wrong. It turns out from the mathematical solution of the Schrödinger equation that in the second shell, the electrons may have orbital angular momentum 1 in addition to orbital angular momentum zero. This fact allows additional possibilities. With orbital angular momentum 1, the component of this angular momentum with respect to an arbitrary axis can have the values 1, 0, or -1. For each of these possibilities, the component of the spin can be either $1/2$ or $-1/2$, for a total of six different states. Therefore, the second shell can contain up to 8 electrons, two with orbital angular momentum 0 and 6 more with orbital angular momentum 1.

After boron comes carbon, with 6 protons in its nucleus, nitrogen with 7 protons, oxygen with 8, and fluorine with 9. The atom with 10 protons is neon, and its electrons fill the first and second shells. After neon comes the atom sodium, with 11 protons in its nucleus and 11 electrons around it. Sodium has one electron in its third shell. This electron has principal quantum number 3. Electrons with this principal quantum number can have orbital angular momentum 0, 1, or 2. Because of the complicated nature of many-body problems, it is difficult to calculate the order in which electrons fill up shells. It turns out that it costs a considerable amount of energy to put an electron in a shell with orbital angular momentum 2, so that an atom with electrons filling all states with orbital angular momentum 0 and 1 acts like an atom with a closed shell. Electrons fill the states with principal quantum number 4 and orbital angular momentum 0 before filling the states with principal quantum number 3 and orbital angular momentum 2. We can go on to consider atoms with more and more protons in their nuclei and more and more surrounding electrons, but the task soon becomes tedious.

The number of protons in the nucleus of an atom is called its "atomic number." The number of nucleons is called the "atomic mass number." The mass of an atom on a scale in which the atom ${}^{12}_{6}C$ (carbon with 6 protons and 6 neutrons) is given the mass 12 on a scale called the "atomic mass." On this scale, the mass of an atom is its mass devided by one twelfth of the mass of ${}^{12}C$, so the atomic mass has no units.

Elements with closed shells have similar properties: namely, they are called "inert gases" because it is hard to make them interact chemically. Elements with one electron outside a closed shell are, except for hydrogen, metals that are quite reactive chemically. Likewise, elements that have two electrons outside a closed shell are chemically similar, and so on.

The elements can be arranged in a table with increasing atomic number from left to right in a row and from top to bottom in a column. In this table, elements with similar chemical properties are put in a column. Such a table is called a "periodic table of the elements." It was first proposed by Dmitri Mendeleev (1834–1907), a Russian chemist. Mendeleev of course lived before anybody knew about quantum mechanics, and so he did not understand the structure of atoms. But he was able to construct the periodic table on empirical grounds, based on their observed chemical properties. The atomic masses of the elements are not whole numbers for two reasons:

1) Protons and neutrons are bound in different nuclei with different binding energies, and the binding energy affects the mass according to Einstein's famous formula $E = mc^2$.

2) A table of atomic masses gives the mass of an element as it occurs most commonly in the earth. The atomic mass of carbon is not exactly 12 because carbon commonly occurs in a mixture of two isotopes, mostly ${}^{12}_{6}\mathrm{C}$ but with a small amount of ${}^{13}_{6}\mathrm{C}$, which is heavier. The symbol for an element with its naturally occurring mixture of isotopes is usually given without the superscript or subscript, for example, C for carbon. Naturally occuring carbon has atomic mass 12.011.

14.3 Compounds

It is no accident that the elements helium and neon, which have closed shells, are inert gases. A molecule of either of these gases consists of a single atom. It is hard (but not impossible) to get inert gases to combine with other elements in order to form compounds. Likewise, it is no accident that the elements lithium and sodium are chemically reactive metals. This is so because each of them has one electron outside a closed shell. This outer electron is called a "valence" electron to distinguish it from the electrons that belong to closed shells. Some elements have more valence electrons than others. For example, carbon has 4 valence electrons, nitrogen 5, and oxygen 6. Valence electrons in an atom are more easily able to combine with other atoms than closed-shell or core electrons.

In the periodic table, neon is directly beneath helium, and sodium is directly beneath lithium. Thus, the position of an element in the periodic table reflects the electron structure of its atoms.

To see why some elements are relatively inert while others react to form chemical compounds, we have to understand the mechanisms that make atoms combine to form molecules. It turns out that it is "energetically favorable" for electrons to form closed shells. In picturesque language, we say that electrons "like" to be in closed shells. Why is this so?

It is a property of atoms and ions that they can exist in different states with different amounts of energy. An atomic state with a high energy can spontaneously radiate light (a photon), which carries off some of its energy. Because of conservation of energy, the atom finds itself in a state of lower energy. This process of the atom emitting photons continues until the atom is in its lowest-energy state, or "ground state." According to quantum mechanics, every atom has a ground state.

Now consider a sodium atom and a fluorine atom placed near each other. If the sodium atom donates its single valence electron to the fluorine atom, the remaining electrons around the sodium will form a closed shell. Similarly, the fluorine atom, by accepting the electron, also has its electrons in a closed shell (it was originally one electron short of a closed shell). After donating an electron, the sodium is no longer a neutral atom but a positively-charged ion. Likewise, after accepting an electron, the fluorine is a negatively charged ion. The positive and negative ions attract each other and form the neutral compound NaF (sodium fluoride). This compound has lower energy than the sum of the energies of the neutral sodium and fluorine atoms. Left to its own devices, the compound is stable, and the chemical bond is called an "ionic" bond. However, the compound can be broken apart if it gains additional energy, for example, by absorbing photons from an external source.

Another type of binding of atoms into molecules is possible. For example, hydrogen gas is not normally a gas of single atoms but a gas of hydrogen molecules. Each molecule is a bound state of two atoms, but one hydrogen atom does not donate its electron to the other. Rather, the two electrons are shared between the two hydrogen nuclei so that in some sense each atom has a closed shell. It turns out that the shared state has lower energy than the energy of two free hydrogen atoms, and therefore is a bound state. This type of chemical bond is called a "covalent" bond.

Atoms forming ionic and covalent bonds are two ways of forming molecules. More complicated, intermediate ways are also possible. It is also possible for atoms to combine into molecules containing three or more atoms. Some complex molecules contain thousands of atoms of many kinds. Particularly complex are some of the so-called organic molecules, which are defined as molecules containing the atom carbon. Because carbon has four valence electrons (four electrons outside its closed shell), it can combine into molecules in many ways with many different kinds of atoms. All life as we know it contains organic molecules, or, in other words, molecules with the element carbon as a constituent.

Some kinds of organic molecules can be synthesized in the laboratory, and it is believed that conditions on earth in its early days were such as to be favorable for the creation of organic compounds. However, life has not been synthesized in the laboratory, and at this time science does not have an answer to the question of how life began.

Chapter 15

Nuclear Physics

> In science there is only physics. All the rest is stamp collecting.
>
> —Lord Ernest Rutherford (1871–1937)

It is ironic that, after putting down other sciences in the remark quoted above, Rutherford won the Nobel prize for chemistry and not for physics. Nevertheless, because he discovered the atomic nucleus, he can be considered the founder of nuclear physics.

15.1 The strong force

As Ernest Rutherford discovered, an atom consists of a tiny central nucleus, in which nearly all the mass of the atom is concentrated, surrounded by one or more electrons. As we have remarked, electrons are believed at present to be elementary but nuclei are complicated, composed of protons and neutrons. Of these two particles, the proton was discovered early in the 20th century and the neutron was not discovered until 1932. Although the subject of nuclear physics began with Rutherford, nuclear physics—the study of the properties of atomic nuclei—came of age only after the discovery of the neutron.

We pointed out in Chapter 11 that an additional force, called the strong force or strong interaction, must exist between protons and neutrons because the electric force is repulsive between protons and almost nonexistent between the neutral neutrons. (Although neutrons are electrically neutral as a whole, they have a distribution of positive and negative charges within them. Neutrons also have magnetism and are subject to magnetic forces.) Thus, the protons and neutrons would not stay bound in a nucleus without the strong force to hold them inside. The strong force between particles in a nucleus is evidently attractive. Protons and neutrons act in a similar way under the strong force, and have masses which differ from each other by just a little over 0.1 percent. Because of their similarity, protons and neutrons are collectively known as nucleons, as we have pointed out in Chapter 11.

We can deduce that the neutron is heavier than the proton, rather than vice versa, because a free neutron decays into a proton plus an electron and antineutrino rather than the proton decaying into a neutron and other particles. This follows because of the law of conservation of energy. The energy of a neutron at rest is $m_n c^2$, where m_n is the mass of the neutron, and, as usual, c is the speed of light. This rest energy must be equal to the sum of the rest energies of the proton, electron, and antineutrino plus the sum of their kinetic energies. Thus, the rest energy of the neutron is greater than the rest energy of the proton, and hence the neutron has a greater mass than the proton.

We know that the electric force between two particles of like charge is repulsive and goes inversely as the square of the distance between them (Coulomb's law). On the other hand, the strong force remains strong while the nucleons are in the nucleus, but if the nucleons are separated by a larger distance, the strong force rapidly weakens and becomes near zero in strength.

To see a consequence of this fact, consider a hydrogen molecule, which is a bound state of two hydrogen atoms. The nucleus of an ordinary hydrogen atom is a single proton. In a hydrogen molecule, the two atoms are about one ten-billionth of a meter (10^{-10} m) apart on the average. This is a sufficient distance that the two nuclei do not feel the strong force, and the binding of the two atoms to form a molecule arises from the electromagnetic force. We call such binding "chemical," as opposed to the "nuclear" binding of nucleons in a nucleus.

It has been determined experimentally that the "range" of the strong force between nucleons is a little over 10^{-15} m. Beyond this range, the force rapidly falls essentially to zero. In 1935, the Japanese physicist Hideki Yukawa (1907–1981) proposed that the strong force between nucleons arises from the exchange of a "meson" field between the nucleons. The quanta of this field are called mesons. From the range of the force Yukawa was able to predict the mass of the mesons.

Yukawa's reasoning goes as follows: There is no action at a distance, so the force between nucleons must arise from a field between the nucleons. If the quanta of the field are massless, like the photon, then the force goes as $1/r^2$. A force that goes like $1/r^2$ is said to have an infinite range. For example, the gravitational force goes like $1/r^2$, and galaxies that are billions of miles away from each other exert gravitational attraction on one another.

However, if the quanta have mass, then the force must have a finite range. The reason is the Heisenberg uncertainty principle. If a nucleon emits a quantum with mass m, then energy is not conserved in this process because at least the rest energy of the emitted particle mc^2 must be created. In the long run, of course, energy must be conserved, but in the short run, energy conservation can be violated by an amount consistent with the Heisenberg uncertainty relation $\Delta E \Delta t \leq \hbar/2$. In the short time Δt the particle can travel at most a distance $c\Delta t$, and that is the range of the force. The larger the mass m of the emitted particle, the shorter the time it can violate conservation of energy and the shorter the range of the force. By knowing the approximate range of the force between nucleons, Yukawa was able to predict the mass of the exchanged meson, using the Heisenberg uncertainty relation. When the meson (now called the pion) was discovered in 1941, its mass was measured to be approximately equal to the mass predicted by Yukawa.

For a number of years, the pion was considered to be the fundamental carrier of the strong force. However, after the suggestion in 1964 by the physicists Murray Gell-Mann and George Zweig (independently) that nucleons and pions are not elementary, but are composed of elementary particles, called "quarks" by Gell-Mann, the picture of the strong force began to change. The picture in 1971 was in terms of a field theory of quarks and massless force carriers called "gluons." We shall see in the next chapter, on the standard model of elementary particles, that the fundamental strong force between quarks is very different in strength and character from the force between nucleons. We shall also see why the range of the force remains short, rather than going like $1/r^2$.

15.2 Atomic nuclei

Under many conditions, the force between nucleons is attractive. It turns out, because of the strength of this attractive force, that if a proton and a neutron are brought close together, within the range of the force between them, they can bind together to form a nucleus called a "deuteron," emitting a photon in the process. The rest energy of the deuteron is less than the sum of the rest energies of the free proton and neutron. The difference is called the "binding energy." In order to free the proton and neutron from each other, an amount of energy greater or equal to the binding energy must be absorbed by the deuteron. If excess energy is absorbed, it can become kinetic energy of the proton and neutron.

Part of the binding of a proton to a neutron to form a deuteron is believed to occur from the exchange of pions between them. If a neutral pion is exchanged, the proton and neutron keep their identities. However, if a positively charged pion is emitted by the proton, it becomes a neutron, and the neutron absorbing the pion becomes a proton. Likewise, a neutron can emit a negatively charged pion, becoming a proton, and the proton absorbing the pion becomes a neutron. We illustrate two of these possible exchanges in Figure 15.1. In the figure, the proton and neutron are depicted as solid lines and the pions as dashed lines. The lines depict the motion of the particles, with space (only one dimension is pictured) plotted from left to right and time plotted upwards. Pictures like this are called Feynman diagrams after the physicist Richard Feynman. Feynman originally used his diagrams to describe the interactions between electrons, positrons, and photons, as we shall see in the next chapter.

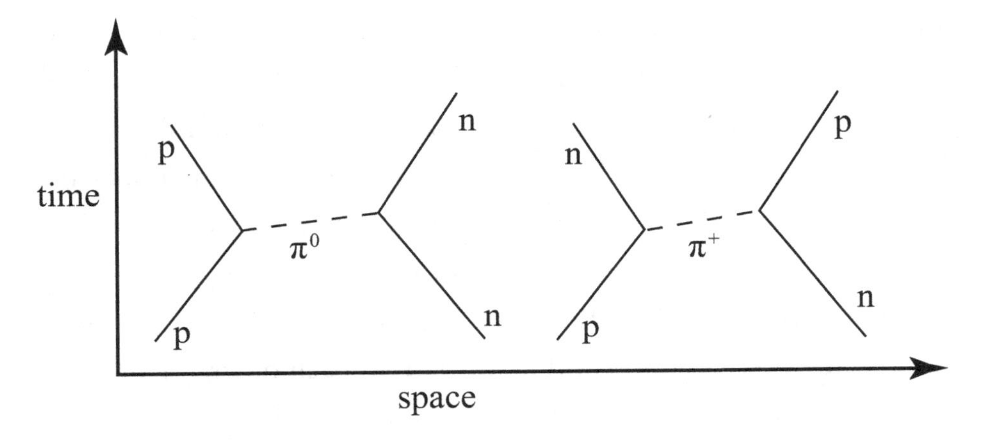

Figure 15.1: Two possible exchanges of a pion between a proton and a neutron. A proton is labeled p, a neutron n, a neutral pion π^0, and a positively charged pion π^+.

A deuteron, like an ordinary hydrogen nucleus, contains one proton. Therefore, a neutral atom containing a deuteron for a nucleus, like an ordinary hydrogen atom, contains one electron outside its nucleus. The atom containing a deuteron for its nucleus is called deuterium or heavy hydrogen, as we have noted in the previous chapter. We have also noted that there exists a still heavier form of hydrogen, containing two neutrons in addition to a proton in its nucleus. This form of hydrogen is called tritium, and its nucleus is called the "triton."

We have remarked in the previous chapter that atoms containing the same number of protons but different numbers of neutrons, are called isotopes. Although different isotopes of the same chemical element have different masses, they have very similar chemical properties. This is because the principal chemical properties of an atom are determined by the number of electrons it contains, which, in the neutral atom, is the same as the number of protons in its nucleus.

As more and more nucleons combine into a nucleus, the binding energy of the nucleus becomes larger and larger, up to a point. Thus, the nuclei of carbon, oxygen, and many other atoms are stable against breaking up into their nucleon constituents. Not only does the binding energy increase as the nuclei become larger, but the binding energy *per nucleon* tends to increase. Because the force between nucleons has complicated properties, the amount of the binding energy is hard to calculate. In most cases, however, it can be measured.

As more and more nucleons are added to a nucleus, the binding energy per nucleon does not continue to increase forever. The reason is the short range of the nuclear force. As a nucleus gains more nucleons, it becomes larger. Consequently, the nucleons on opposite edges of a large nucleus are too far apart to feel the effects of the strong force. However, the range of the repulsive electric force (Coulomb force) between protons in the nucleus is infinite, so that the repulsive force is felt among *all* protons in the nucleus. Eventually, for large enough nuclei, the repulsive electric force, although intrinsically much weaker than the attractive nuclear force, wins out because of its long range. This fact limits the number of protons in a nucleus.

You might wonder at this stage why we don't have nuclei with an arbitrarily large number of neutrons. The reason has to do with the details of the nuclear force. The strength of the force between two nucleons depends on the quantum state of the two particles. It turns out certain quantum states that are allowed for a proton and neutron are forbidden by the Pauli principle for two protons or two neutrons. It is just in a state forbidden for two protons or two neutrons that the force between a proton and neutron is the strongest and leads to the binding of the proton and neutron. The force is not strong enough to bind a nucleus with only neutrons or only protons in states that are allowed for them. For relatively small nuclei, such as carbon or oxygen, the most stable nuclei contain equal numbers of protons and neutrons. Large nuclei, such as gold, have more neutrons than protons because the repulsive electric force between protons reduces the number of protons relative to the number of neutrons.

The result of all this is that the binding energy per nucleon increases with increasing number of nucleons up to a certain point, which happens to be ^{56}Fe, an isotope of iron. Then, for still larger nuclei, the binding energy per nucleon decreases as the number of nucleons increase further. As the number of nucleons in a nucleus continues to increase, eventually the nucleus becomes unstable and decays into a nucleus containing a smaller number of nucleons. In these decays, the energy, charge, and number of nucleons are all conserved.

15.3 Nuclear fusion

The sun consists mostly of the element hydrogen in the isotope ^{1}H. In the core of the sun most hydrogen nuclei are separated from their electrons, and wander about as protons. Under proper conditions involving multistep processes, four of these protons can fuse together to form a helium nucleus, which is a bound state of two protons and two neutrons.

A large amount of energy is released in these fusion processes. Both the strong and the weak interactions play a roll, the weak interaction being involved whenever a neutrino is emitted. The result of the fusion of hydrogen into helium is the loss of less than 1 per cent of the mass of the original four protons. This small amount of mass is converted into a large amount of energy, according to Einstein's equation $E = mc^2$.

The two released positrons in these fusion reactions annihilate against two electrons via the electromagnetic interaction, producing energetic photons. The photons are absorbed in the sun, and less energetic photons are emitted. Many absorption and reemission processes take place, degrading the energies of the photons. Eventually, most of the electromagnetic energy is radiated out into empty space, much of it in the form of visible light. The neutrinos escape from the sun with very little interaction and and are also radiated out into space. Many of the neutrinos reach the earth and some of them can be detected on earth by large scientific instruments. Our eyes are not sensitive to neutrinos, and so we cannot "see" them. Billions of neutrinos from the sun enter our bodies every second. But because neutrinos interact very weakly, nearly all of the neutrinos incident on our bodies go right through them with no interaction at all. In fact, most neutrinos impinging on the earth go right through it without interacting.

What are the proper conditions for the fusion of hydrogen into helium? The protons must have a large kinetic energy to overcome the repulsive electrical interaction due to their like charges and must approach each other within the range of the strong and weak forces. These conditions are met in the core region of stars, including our sun. In fact, the fusion of hydrogen into helium is the main reaction that produces the energy from the sun. Other fusion reactions play a smaller role in the sun.

The nuclear fusion processes take place in the central core of the sun, which has a temperature of about 10 million K at the center. (This is about 18 million degrees Fahrenheit.) Nuclear fusion also takes place in stars, which, like the sun, shine from the electromagnetic radiation released as a consequence of nuclear fusion.

Much effort has been devoted to trying to obtain controlled energy from fusion reactions on earth. So far, there has been no economic success, although fusion has been made to occur. We have achieved uncontrolled fusion in the hydrogen bomb.

One of the main efforts so far to achieve controlled nuclear fusion has been to fuse two tritium nuclei together in a "fusion reactor", according to the reaction

$$^3H +^3 H \rightarrow^4 He + 2n,$$

where H and He are the chemical symbols for hydrogen and helium respectively, the left superscript denotes the number of nucleons (protons plus neutrons) in the nucleus, and n is the symbol for a neutron. More than a million times as much energy is released in this nuclear reaction than in a typical chemical reaction.

Hydrogen gas at ordinary temperatures does not undergo fusion. The reason is that the hydrogen nuclei do not have nearly enough kinetic energy to overcome the repulsive Coulomb interaction between them and approach each other sufficiently closely to allow the short-range nuclear force to be active and cause them to fuse. The hydrogen must be heated to a sufficiently high temperature for fusion to occur, and that temperature is millions of degrees. No container made of matter can withstand such high temperatures. In the sun, gravity holds the material together, but gravity is a negligible force in a fusion reactor on earth. So on earth the principal attempt has been to heat the hydrogen to sufficiently high temperature that it ionizes, or, in other words, the electrons are set free from the nuclei. An ionized gas is called a "plasma."

In order to heat the plasma further, it is necessary to prevent the particles from hitting the walls of the container, which would vaporize if struck by many of the ions. But because the ions are charged particles, their paths may be controlled by a magnetic field. The problem has been, first, to heat the particles to a sufficiently high temperature, and second, to get the particles to circulate in the magnetic field for a sufficiently long time that they strike one another and fuse. A Small amount of fusion has already been achieved, but it may be a long while before we achieve enough fusion in a fusion reactor to make the reactor economically feasible.

In a hydrogen bomb, we do not need to worry about vaporizing the container. As soon as the uncontrolled fusion takes place, the container is vaporized, but that is a desired result of the nuclear explosion. In a hydrogen bomb, the explosion of an ordinary atom bomb in the same container as the hydrogen, heats the hydrogen sufficiently to cause fusion. This "ordinary" atom bomb is a "fission" bomb, which we shall describe two sections later.

15.4 More about radioactivity

The discoverer of radioactivity was Antoine Henri Becquerel (1852–1908), a French physicist. He discovered in 1896 that a compound of uranium fogged a photographic plate in the dark. He correctly deduced that something emitted by the uranium was responsible. The effect was called radioactivity.

What is the reason for radioactivity? We have remarked that there is a limit to the number of nucleons in atomic nuclei. The reason, as we have noted, is that the protons in a nucleus are subject to a repulsive electral force (the Coulomb force) that has an infinite range and so as the number of protons increase, eventually the repulsive electrical force becomes more effective than the attractive nuclear force, with its short range. Furthermore, the nature of the strong force is such that the number of neutrons in a nucleus cannot greatly exceed the number of protons, so we cannot have a nucleus with, say, 100 protons and an arbitrarily large number of neutrons.

When a nucleus becomes heavy (that is, contains substantially more than 200 nucleons, it becomes energetically favorable for the nucleus to emit radiation, becoming a lighter nucleus. For example, the nucleus $^{238}_{92}U$, the most common isotope of uranium, can spontaneously decay by emitting an α particle (a helium nucleus containing two protons and two neutrons), becoming $^{234}_{90}Th$ (the element thorium). We say the uranium is the parent nucleus and the thorium is the daughter nucleus. The half-life of ^{238}U is about 4.5 billion years (4.5×10^9 years), which by coincidence is about the same as the age of the earth.

The thorium is itself unstable, and decays via β decay into still another element. This beta decay occurs via the weak interaction, but the probailities are such that the half-life of the thorium is only 24 days. Thus, we see that the strength of an interaction is only one of a number of factors that determine the lifetime of an unstable element. The decay product of thorium is also unstable. In a series of α and β decays, the original uranium eventually becomes a stable isotope of lead. We see that natural radioactivity by nuclei is a way in which one element can be transmuted into another—a sort of natural alchemy.

Just by looking at uranium and its decay products, we are able to account for natural radioactivity of α and β rays. What about γ rays? A γ ray is an energetic photon emitted by a radioactive nucleus. It doesn't change one kind of nucleus into another. Sometimes, after a radioactive decay, the daughter nucleus is left in an excited state (a state with more energy than its lowest-energy state, or ground state). If that occurs, the daughter nucleus may emit a γ ray, going into a state of lower energy, which may or may not be the ground state. The emission of γ rays goes via the electromagnetic interaction.

15.5 Nuclear fission

Although radiactive nuclei usually decay by α, β, or γ emission, some nuclei occasionally decay by breaking up into two lighter nuclei. We call this breakup "spontaneous fission." An isotope of uranium, ^{235}U, which has three fewer neutrons than the more common ^{238}U, sometimes undergoes spontaneous fission. When a nucleus like ^{235}U undergoes spontaneous fission, not only are two smaller nuclei emitted but also one or more free neutrons. If one of these free neutron strikes another ^{235}U, the neutron can cause it to undergo what is called "induced" nuclear fission. In a fission process, the energy released is more than a million times greater than in a chemical reaction between two atoms.

Naturally occuring uranium, as mined from the ground, contains a mixture of more than 99 per cent ^{238}U and less than one per cent ^{235}U. Because the ^{235}U is so rare, free neutrons emitted by a ^{235}U nucleus are much more likely to strike a ^{238}U than another ^{235}U. However, if ^{235}U is somehow separated from ^{238}U, in a process known as "enrichment," the neutrons emitted in the spontaneous fission can induce fission in the rest of the material, in what is known as a "chain reaction." Such a chain reaction can lead to a nuclear explosion, in what is known as an atom bomb.

A small amount of ^{235}U does not spontaneously explode because many of the neutrons reach the surface of the material and escape before they can induce fission. In order to undergo an explosion, there must be a sufficient amount of the material, called the "critical mass," so that most of the neutrons do not escape but induce nuclear fission in other nuclei.

The essence of a uranium atom bomb is that the highly enriched uranium is initially divided into two pieces, each of which is sub-critical and so does not explode. The bomb is detonated by an ordinary chemical explosive, which drives the two pieces of uranium together into a single piece that is super-critical. The nuclear explosion then naturally occurs.

Because ^{238}U and ^{235}U are different isotopes of the same chemical element, they cannot be separated by chemical means. Therefore, they are separated by physical processes that take advantage of the different masses of the two isotopes. One method is gaseous diffusion. The uranium is combined with another element into a gas, and the gas is allowed to diffuse through a porous material. The lighter gas diffuses more rapidly than the heavier gas, which is left behind. This diffusion process must be repeated many times, each time leading to a slight enrichment of the lighter material. Another method is to use a centrifuge, which spins the material. The heavier gas gets spun preferentially to the outside, leaving the enriched material behind. Again this process has to be repeated many times.

Uranium is not always exploded but may also be "burned" in a controllable way in a nuclear reactor. A reactor takes advantage of the fact that if the neutrons emitted in a spontaneous fission are slowed down, the slow neutrons can make ^{238}U also undergo induced fission. So a substance, called a "moderator," is introduced into a reactor to slow down the neutrons. Sometimes the moderator is heavy water (water with deuterium instead of ordinary hydrogen combined with oxygen), and sometimes the moderator is graphite (a form of carbon). To prevent a reactor from undergoing an uncontrolled chain reaction leading to a nuclear explosion, an essential part of a reactor is a method to control the number of neutrons hitting the uranium. This task is accomplished with control rods made of a substance that absorbs neutrons. If the control rods are inserted more deeply into the reactor, the nuclear reactions are slowed, and if the rods are partially removed, the nuclear reactions are speeded up. The depth of the rods is adjusted until the reaction is just critical, and the reaction is controlled. Heat is removed from the reactor and is used to generate electricity just as heat from a coal or oil generator does.

One of the processes that occur in a nuclear reactor is the transmutation of uranium into plutonium. One isotope of plutonium, $^{239}_{94}\mathrm{Pu}$, is itself a fissionable material, and can be used either in reactors or in atomic bombs. Plutonium, being a different chemical element from uranium, can be separated from it chemically. Because of certain physical properties of plutonium, a plutonium bomb cannot be made by taking two pieces of the substance and making them collide. Rather, a hollow sphere of plutonium is made, and chemical explosives are put around the sphere. When detonated, the plutonium is "imploded" into a supercritical mass, and a nuclear explosion occurs.

During World War II, the United States dropped two atom bombs on Japan. A uranium bomb destroyed the Japanese city of Hiroshima on August 6, 1945, and a plutonium bomb destroyed the city of Nagasaki three days later. Together, these bombs killed about 200,000 people, most of them women, children, and old people, as most of the young men were away in the Japanese armed forces. Although there have been many tests of both fission bombs (atom bombs) and the much more powerful fusion bombs (hydrogen bombs) since then, as of the time of this writing, no other nuclear weapons have been intentionally used on people.

Chapter 16

Elementary Particles

—Three quarks for Muster Mark!

—James Joyce (1882–1941) in *Finnegans Wake*

16.1 What is an elementary particle?

Whether an object or a particle is elementary or composite can be answered only in the context of a theory. For example, in the context of Newtonian gravity, the earth and sun can to a good approximation be regarded as elementary point-like particles whose mutual attraction causes the earth to go around the sun in an ellipse. Of course, in asking other questions, such as what makes the sun shine or why there are tides on earth, we have to regard the sun and earth as composite objects.

As another example, this one on a microscopic scale, a dilute gas can be considered to be made up of essentially point-like molecules, whose structure does not affect the gross properties of the gas. But if we go deeper into the structure of molecules, we learn that they are composite particles.

At the present time, our best theory of what are really elementary particles is the so-called "standard model of elementary particles" or just "standard model" for short. Why it is called a model rather than a theory is partly historical and partly because most physicists do not think it is the last word on the subject.

According to the standard model, molecules are composite, being made of atoms. Atoms are themselves composite, made of electrons and atomic nuclei. In turn, nuclei are composite, being composed of protons and neutrons. Finally, protons and neutrons are composite, being made of quarks, which are elementary according to the standard model. Electrons are also elementary in the model.

But electrons and quarks are not the only elementary particles in the standard model. Before we enumerate them all, we discuss developments leading to the standard model.

16.2 The fundamental strong force

In the previous chapter, we discussed the strong force between nucleons (protons and neutrons). But nucleons are not elementary particles according to our present best understanding. Rather, they are composite, being made up of elementary particles called quarks. Gell-Mann took the name quark from *Finnegans Wake*, a novel by James Joyce. Gell-Mann used the name because in Joyce's novel there are three quarks ("three quarks for Muster Mark"), and in Gell-Mann's original model, there were only three different kinds of quarks. Now we know that there are more kinds of quarks, but the number three still is important in the theory. For example, in an oversimplified version of the modern theory, a nucleon (proton or neutron) contains three quarks.

The quarks in a nucleon are of two kinds, whimsically called "up" and "down," with symbols u and d respectively. The proton is made of two up quarks and a down quark, "uud," and the neutron of two down quarks and an up quark, "ddu." The fact that the neutron is heavier than the proton is attributed to the fact that the d quark is heavier than the u quark. At the present time we have no understanding of why the d quark has a larger mass than the u quark. At present, it is just an experimental fact, obtained by indirect measurements. (The measurements have to be indirect, because so far we have not been able to observe a free quark in order to make a direct measurement of its mass.)

A meson, such as a pion, is made of one quark and one antiquark. Pions exist in three states of electric charge. A π^+ is $u\bar{d}$, a π^- is $d\bar{u}$, and a π^0 is an equal combination of $u\bar{u}$ and $d\bar{d}$. The π^+ and π^- are antiparticles of each other, and the π^0 is its own antiparticle. Pions are not stable but decay rapidly into lighter particles. The π^+ most commonly decays into $\mu^+ + \nu_\mu$, with a mean life of 2.6×10^{-8} s, where the μ^+ is a charged particle called the "muon," and the ν_μ is a neutrino associated with the muon. We discuss these particles further in the next section. The π^- most commonly decays into the $\mu^- + \bar{\nu}_\mu$ with the same lifetime. The π^0 decays into two photons with a lifetime of 0.8×10^{-16} s.

In our present understanding, the fundamental strong force is not between nucleons or between nucleons and pions but between quarks. This force is so strong that quarks are confined inside nucleons or mesons. This is why the strong force has an effective short range, even though the carrier of the strong force is believed to have mass zero. The force between nucleons or between a nucleon and a meson is only a residual effect of the fundamental strong force, as we shall discuss in more detail in section 4.

16.3 The weak force

We have pointed out that the neutron and proton are not elementary particles but are composites of quarks. In the quark picture, a neutron decays into a proton because one of the d quarks inside the neutron decays into a u quark via the weak force. The interaction is

$$d \rightarrow u + e^- + \bar{\nu}_e.$$

In the above interaction, we have put a subscript e on the symbol for the antineutrino to indicate that it is an antineutrino associated with an electron. By this we mean that if an electron and a neutrino are created together in an interaction, the neutrino will be an electron-type neutrino.

We point out here and in the next section that actually three different kinds of neutrino are known, one of which is associated with the electron, and the other two associated with charged particles analogous to the electron but much heavier. These heavier charged particles are known as the muon (pronounced mu-on), symbol μ^- and the tauon (rhymes with now on), symbol τ^-. The tauon is often called the "tau lepton." The muon and tauon are unstable, decaying via the weak interaction. The muon decays by

$$\mu^- \to e^- + \bar{\nu}_e + \nu_\mu.$$

Note that electric charge is conserved in the decay. Also note that the two neutrinos are emitted in the decay, one being an electron-type (anti)neutrino and the other being a muon-type neutrino. The average lifetime of the muon is only about 2 millionths of a second (2×10^{-6} s), which is much shorter than that of the neutron. Yet both of these lifetimes can be approximately calculated with the weak interaction. The lifetime of the tauon is much shorter than that of the muon. In fact, the muon's lifetime, short as it is, is about 10 million times as long as the lifetime of the tauon.

It may seem strange that three different particles, the n, the μ, and the τ, have such different lifetimes, when they all decay because of the weak interaction. The explanation is that there are other factors (which we discuss shortly) besides the kind of interaction that govern the lifetime of a decaying particle. Different decaying particles decay into different final particles.

One of the most important factors governing the lifetime of a decaying particle is the amount of kinetic energy released in the decay. If the sum of the masses of the final particles is only a little smaller than the mass of the initial particle, as in the case of the neutron, the final particles will have very little kinetic energy. All other things being equal, the less kinetic energy available in the final state, the longer will be the lifetime. This fact follows from the rules of calculation in quantum mechanics. We can make this calculational rule somewhat plausible by noting that if the sum of the masses of the final particles is greater than the mass of the initial particle, the initial particle cannot decay at all because energy would not be conserved. So as the kinetic energy of the final particles gets smaller and smaller, the lifetime gets longer and longer, until, when the kinetic energy is zero, the lifetime becomes infinite, or, in other words, the decay does not happen. There are no cases that we know of in which the sum of the masses of the potential decay products is exactly the same as the mass of the potential decaying particle. The sum of the masses of the final particles is either less than that of the initial particle, in which case the decay occurs, with the final particles having kinetic energy, or the sum of the masses of the potential final particles is greater than the mass of the initial particle, in which no decay into that final state occurs.

The three charged particles, e^-, μ^-, and τ^-, and the three neutral particles ν_e, ν_μ, and ν_τ, are called "leptons." Thus, we know of the existence of three charged leptons and three neutral leptons, for a total of six in all. There are also three distinct charged antileptons, but we do not yet know whether the three neutral antileptons are the same as or different from the three neutral leptons.

Just as we know of six different kinds of leptons, we know of six different kinds of quarks. They are called the up (u), down (d), strange (s), charmed (c), bottom (b), and top (t) quarks, in order of increasing mass. The names are whimsical and historical, and have nothing to do with the properties of the quarks. All the quarks and all the charged leptons participate in the electromagnetic interaction. The quarks and the leptons all participate in the weak interaction. However the quarks, but not the leptons, also participate in the strong interaction. It is known that the range of the weak interaction is even smaller than the range of the strong interaction.

We shall see in the next section that the weak, strong, and electromagnetic interactions are built into the framework of the standard model.

16.4 The standard model

The standard model is a quantum field theory, which means that the particles of the theory are quanta of fields. The model includes quantum electrodynamics, in which photons are quanta of the electromagnetic field and electrons (and their antiparticles) are quanta of the electron field. But these fields are only two of the quantum fields that make up the standard model. There are also quantum fields corresponding to the strong and weak forces, as well as quanta of other matter fields besides that of the electron. We discuss these fields later in this section.

Gravity is outside the framework of the standard model, but presumably there exists a quantum gravitational field. So far, however, there have been difficulties in constructing a quantum theory of gravity, and there is not yet a generally accepted theory. The present best theory of gravity is Einstein's theory of general relativity, but it is a classical theory in the sense that it is not quantized. There is a problem of consistency if the theory of gravity is not quantized while the theory of all the other interactions is quantized, so attempts have been made to modify general relativity to a form that can be consistently quantized. We briefly discuss one of these attempts in the last chapter.

Among the force fields of the standard model is the electromagnetic field, whose quanta are the photons. The electromagnetic field carries the force between electrically charged particles, such as electrons. The theory requires photons to have spin one, and that is the value of their measured spins. Photons have zero masses in the model, and, experimentally, to the best of our measurements, photon masses are zero. There is only one kind of photon, but photons come in a wide range of energies (or frequencies). Recall that the energy E of a photon is proportional to its frequency f according to the formula $E = hf$, where h is Planck's constant.

The theory of the elecromagnetic field interacting with charged particles is called "quantum electrodynamics" because it is a a dynamical theory, dealing with forces (i.e. interactions) and motion and because the fields are quantized into particles. In Figure 16.1 we show Feynman diagrams of electrons and positrons interacting with photons. In (a) two electrons scatter because they exchange a photon. In (b) an electron and a positron annihilate with the creation of two photons. We use e^- as the symbol for an electron, e^+ as the symbol for a positron, and γ as the symbol for a photon.

Among the particles subject to the strong force are the proton and neutron, each of which has spin 1/2. Both are constituents of atomic nuclei. In the standard model, the proton and neutron are are composite bound states, each of which contains three elementary particles called quarks. The quarks are matter fields and have spin 1/2. According to the rules of addition of spin in quantum mechanics, if an odd number of spin 1/2 particles are bound together, the resulting

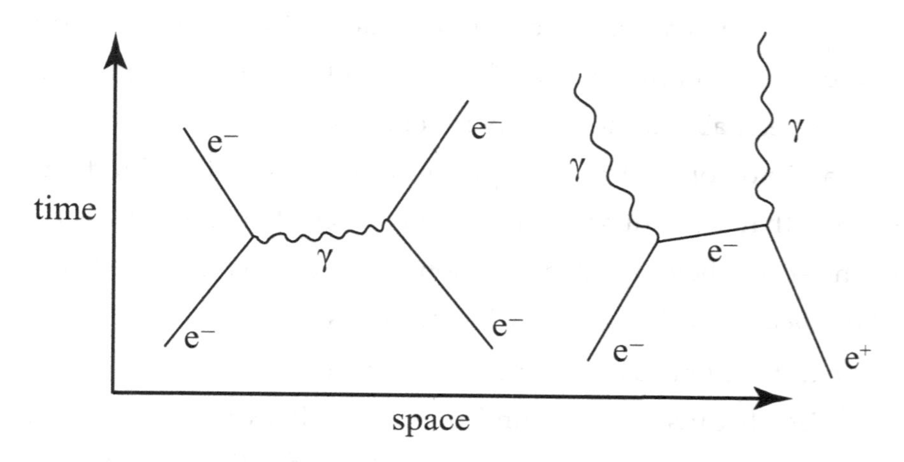

Figure 16.1: Electron and positron interactions with photons. In (a) we show two electrons scattering by exchanging a photon. In (b) we show an electron and positron annihilating into two photons. Electrons and positrons are denoted by straight lines and photons by wiggly lines.

composite particle must have half-integral spin. If an even number of spin 1/2 particles are bound together, the resulting particle must have integer spin. The forces among the three quarks in a proton or neutron are such that the proton and neutron each have spin 1/2. In principle they could have a higher spin (spin 3/2 is allowed by the rules of quantum mechanics, but, as we said, the forces actually lead to spin 1/2), but they could not have spin zero or one.

Just as the spin-1 electromagnetic field is responsible for the electromagnetic force between charged particles, spin-1 strong fields are responsible for the strong force between quarks. There is, however, a qualitative difference between the strong and electromagnetic field. If enough energy is supplied to an electromagnetically bound state, like the hydrogen atom, the electron is knocked loose from the proton and can travel freely away from it. On the other hand, nobody has ever been able to do an experiment that knocks a single quark out of a proton or neutron, and physicists believe that the binding is so strong that a free quark cannot be observed. Perhaps for this reason, the strong field is called a "gluon" field, because it is the "glue" that binds the quarks tightly inside nucleons.

The different behavior of the strong and electromagnetic interactions arises because of profound differences between photons (the quanta of the electromagnetic field) and gluons (the quanta of the strong field). The photons interact with all electrically charged particles, but are themselves electrically neutral. Therefore, photons interact with each other only indirectly. A photon can convert to charged particles, such as an electron and positron pair, and another photon can interact with either the electron or positron before the pair annihilate back into a photon. This complicated process, which is allowed in quantum field theory, leads to an indirect interaction between two photons.

However, the gluons interact with each other more directly. The reason is that gluons interact with all particles, like quarks, that carry strong charge, and the gluons themselves have strong charge. Partly for whimsical reasons, strong charge has been called "color." The term has nothing to do with what we ordinarily mean by color. Whereas there is only one kind of electric charge (and its anticharge, which is opposite in sign), there are three kinds of strong charge or color, which have been denoted "red," "green," and "blue." Perhaps a better way of describing these strong charges would be to call them "strong-charge-1," "strong-charge-2," and "strong-charge-3," but we shall use the more common designation of three colors. There are also three anticolors.

The theory says that a quark, say the u quark, comes in three colors. Thus, there exist a red u quark, a green u quark, and a blue u quark. Likewise, there exist red, green, and blue d quarks. Because this theory of strong interactions of colored particles is analogous to the theory of quantum electrodynamics, the strong theory is called "quantum chromodynamics" (chromo for color).

Just as in electromagnetism, an electrically charged particle can annihilate with its antipartice (for example, an electron and a postiron can annihilate) to produce a photon, so a colored particle, say a u quark, can annihilate with an anti-u quark ($\bar{u}$ quark) to produce a gluon. It is this annihilation process that illustrates the distinction between photons and gluons.

Consider a photon created in the annihilation of an electron and a positron. Because electric charge is conserved in the theory, the photon must be electrically neutral because the opposite charges of the electron and positron cancel in the annihilation.

Now consider a gluon created in the annihilation of a u quark and a $\bar{u}$ quark. Just as electric charge is conserved, so is color. Now suppose that a red u quark annihilates against an antigreen $\bar{u}$ antiquark. Then the created gluon cannot be color-neutral but must have the colors red-antigreen. Combining three colors and three anticolors, we get nine possibilities. Therefore, naively we might expect that there exist nine different kinds of gluons. This expectation is not quite right. One of the nine possibilities turns out to correspond to a color-neutral combination, and so is not a color gluon. In the theory, there are eight different kinds of gluons, each of which has color (and anticolor). Because, in the theory, gluons interact with all colored particles, gluons interact directly with one another.

In Figure 16.2 we show Feynman diagrams comparing (a) the annihilation of an electron and positron to create a single photon with (b) the annihilation of a quark and antiquark to create a single gluon. The annihilation shown in (a) cannot take place in free space because it would be impossible to conserve both energy and momentum, but it can occur in the neighborhood of another charged particle. The annihilation shown in (b) also does not occur in free space, not only because it would be impossible to conserve energy and momentum but also because quarks and gluons are confined to the interior of strongly interacting particles such as protons and pions. In (a), because the electron and positron are oppositely charged with the same magnitude of the charge, the photon must be electrically neutral to conserve electric charge. In (b), because the quark is blue (for example) and the antiquark green, the gluon must be blue-antigreen in order to conserve color.

We have remarked that the proton and neutron each is a composite particle of three quarks. One of these quarks is red, one is green, and one is blue, and the wave function is such that the proton and neutron are color-neutral combinations of the three colored quarks. It is believed, although not rigorously proved from the theory, that only color-neutral particles can exist as free particles. Quarks and gluons, having color, are tightly bound in the interior of the color-neutral particles. This is the reason why the strong force between

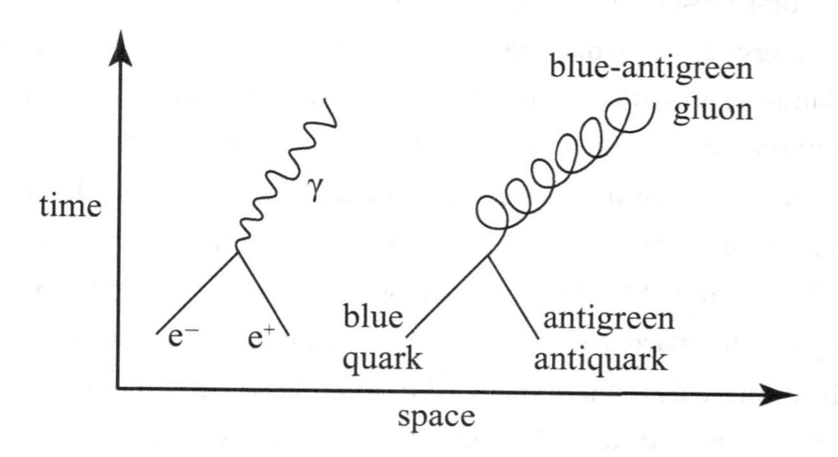

Figure 16.2: (a) Annihilation of an electron and positron into a single electrically neutral photon. (b) Annihilation of a blue quark and an antigreen antiquark into a single blue-antigreen gluon. Neither interaction can take place in free space. It is usual to depict a gluon schematically as a spring, and we have done so here.

quarks does not go as $1/r^2$ even though the gluons are massless, according to the theory.

Color-neutral particles composed of quarks, antiquarks, and/or gluons, are called "hadrons." The proton and neutron are just two examples of hadrons, although they are very important because they are the constituents of atomic nuclei. Hadrons that contain three quarks, like the proton and neutron, are called "baryons," while hadrons that contain a quark and an antiquark are called "mesons." If a particle exists, then quantum field theory says that its antiparticle can also exist. Since baryons and mesons exist, antibaryons and antimesons must also have the possibility of existing. In fact, the antiparticles of many observed hadrons have been created in the laboratory.

The first meson to be discovered was the so-called pi meson or pion, observed in cosmic rays in 1941. Cosmic rays consist of various kinds of radiation from outer space, which strike the earth's atmosphere. Pions, however, have very short average lifetimes, so they cannot come from outer space. Rather, they are produced in the atmosphere in collisions of primary cosmic rays with nuclei of the atmosphere. Many of the primary cosmic rays are protons, but a large variety of other particles also impinge on the earth's atmosphere.

The pion exists in three states with different electric charges: positive, negative, and neutral. The pions have masses that are about 1/7-th of the mass of the proton. Since the discovery of the pion, many other kinds of mesons and baryons have been discovered. All mesons are unstable, and all baryons except the proton are unstable. The ultimate products of the decays are stable particles, which include the proton, electron, photon, and neutrino. The stable particles are not necessarily completely stable, but if they decay, their average lifetimes are so long that their decays have not yet been observed. If the proton is unstable, its average lifetime is much longer than the age of the universe, which is about 14 billion years. There are so many protons in the human body that if the proton had a lifetime no longer than the age of the universe, the radioactivity from proton decay would prevent us from living.

Why is the proton so stable? Physicists like to have a reason for things. Murray Gell-Mann likes to state the "totalitarian principle" of physics: "Whatever is not forbidden is compulsory." If the proton does not decay, there must be a reason, but we don't know the reason. Anyway, physicists have invented a conservation law to account for the lack of observed decays of the proton. This law is called the conservation of baryon number. The law says that the number of baryons minus the number of antibaryons remains constant as time goes on. There is reason to think that the law of baryon number conservation is not exact and was violated under conditions of the early universe so that our universe at present can contain many more baryons than antibaryons.

A proton and a neutron are baryons, and they are each assigned baryon number equal to 1. Notice when a neutron decays

$$n \to p + e^- + \bar{\nu}_e,$$

a baryon (the neutron) is destroyed but another baryon (the proton) is created so that the baryon number remains constant at unity.

We next turn to the weak interactions. The force carriers of this interaction are called "weak bosons." There is an electrically charged weak boson called the W^+ and its antiparticle, the W^-, as well as an electrically neutral weak boson called the Z. The Z, like the photon, is identical to its antiparticle. Unlike the photon and the gluons, which have zero mass, the W and Z bosons are quite heavy; the W has a mass about 85 times as large as the mass of the proton, and the Z has a mass about 97 times as large as that of the proton.

The large masses of the W and Z are responsible for the fact that the weak interaction appears weak. According to quantum field theory, the weak interaction between two matter particles is transmitted by a W or Z boson traveling from one particle to the other. But if a W or Z is emitted by one particle, energy is not conserved, as at least mc^2 of energy must be created to produce the W or Z, where m is the mass of the W or Z. Although energy is conserved on the whole, the Heisenberg uncertainty principle allows for a violation of energy conservation for a small time t, so long as the energy E created times the time of the violation satisfies the uncertainty relation $Et = \hbar/2$. But when the W or Z is produced, the energy is large because the mass is large (remember $E = mc^2$). Because E is large, t is very small. In the very small time the W or Z are traveling between matter particles, these bosons cannot go very far. That means that the interaction is effective only when the two matter particles are very close together. Since that happens only rarely, the interaction occurs only rarely, and the interaction appears weak. Thus, the apparent weakness of the so-called weak interaction arises because of its short range.

Among the particles of matter we have considered are the u and d quarks and the electron e^- and its neutrino ν_e. Except for the ν_e, these are the particles that are constituents of ordinary atoms. The ν_e is emitted (actually an antineutrino) when some unstable atomic nuclei decay. When an unstable nucleus decays by emitting an electron and an antineutrino, the fundamental process is that a d quark in one of the neutrons in the nucleus turns into a u quark, emitting a W^- boson. The u quark remains bound in what was formerly a neutron (with constituents ddu) but becomes a proton (with constituents udu). The W^- boson, in turn, disappears, creating an electron and a $\bar{\nu}_e$. We illustrate this fundamental process in in Figure 16.3.

In the standard model, the four particles u, d, e^-, and ν_e belong to a a collection of elementary particles called a "family." In the stan-

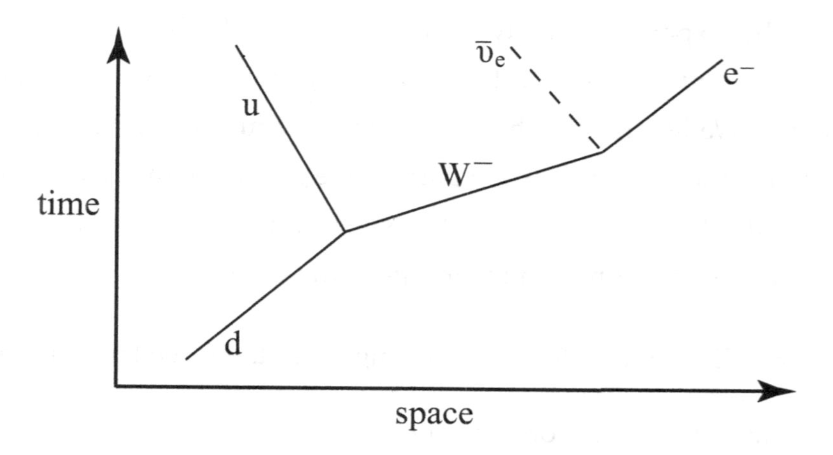

Figure 16.3: Decay of a d quark into a u quark, electron e^-, and antielectron neutrino $\bar{\nu}_e$ with an intermediate weak boson W^-. A neutrino or antineutrino is denoted by a dashed line.

dard model as we know it at present, there exist two other families of particles, for a total of three. Each family consists of two different kinds of leptons and two different kinds of quarks (each of which comes in three colors). These different kinds of particles are often called "flavors." Leptons are spin 1/2 particles that do not have strong interactions but have weak interactions, and, if electrically charged, have electromagnetic interactions. The two leptons of the first family are the negatively charged electron and an accompanying electrically neutral neutrino, called the electron neutrino. The electric charge of the u quark is 2/3 in units of the charge of the proton, and the electric charge of the d quark is $-1/3$ in the same units.

The second and third families each consist of two quarks and two leptons with the same pattern of electric charges as the first family. Why are there three families? The standard model does not say, but that is what experimental physicists have found. There could in principle be more families, but if so, there is at present no hint of their existence. We list the members of the three families in Table 16.1. In addition to the symbol for each particle we give its name. The names were partly chosen for whimsical reasons, and have essentially nothing to do with the common meanings of the words.

Table 16.1: Elementary fermions belonging to the three known families

First family	Second family	Third family
ν_e (e-neutrino)	ν_μ (μ-neutrino)	ν_τ (τ-neutrino)
e (electron)	μ (muon)	τ (tau lepton)
u (up quark	c (charmed quark)	t (top quark)
d (down quark)	s (strange quark)	b (bottom quark)

In Table 16.1 we have omitted color. Because each quark can be red, green, or blue, each family actually contains six quarks and two leptons. Each of the charged fermions has an antiparticle with the opposite electric charge. The neutrinos are electrically neutral, and, as we have said, it is not yet known whether antineutrinos are the same or different from neutrinos.

What are the differences between analogous members of different families? In the first place, their masses are different. In the second place, their flavors are different, so we have in all six quark flavors and six lepton flavors. Flavor appears to be conserved in strong and electromagnetic interactions, but not in weak interactions. This means, for example, that a c quark cannot decay into a u quark via the strong or electromagnetic interactions. Furthermore, it seems as if the c quark cannot decay into a u quark even via the weak interaction. However, the weak interaction allows the c quark to decay into an s or d quark plus other particles via the weak interaction. The standard model is constructed so as to allow these decays, and they have been indirectly observed experimentally. Actually, what is directly observed is a hadron containing, for example, a c quark decay into a hadron containing a d or s quark. Similarly, a hadron containing a d quark can decay into a hadron containing a u quark. As we have seen, this is the mechanism for a neutron to decay into a proton.

It is interesting to say something about the masses of the quarks and leptons. In the study of elementary particle physics, it is customary to give the rest energies of these particles, using as a fundamental unit the "electron volt," abbreviated eV. An electron volt of energy is defined to be the energy acquired by an electron when it is accelerated by one volt of electric potential. Secondary units often used are million electron volts (MeV) and billion electron volts (GeV). As an example, an electron, which has a mass of 9.11×10^{-31} kg, has a rest energy of 0.511 MeV. As another example, a proton, which has a mass of 1.67×10^{-27} kg, has a rest energy of 938 MeV or 0.938 GeV. These examples show that the eV is a very tiny unit of energy.

The masses of the charged leptons and the masses of some hadrons, such as the proton and neutron, have been measured very precisely. The same cannot be said for quarks and neutrinos, for different reasons. First, we consider quarks. In order to measure a mass of a particle precisely, we have to measure it when it is free. Because quarks are always bound in the interior of hadrons, with the exception of the t quark, which we consider separately, we have no way to measure quark masses directly. Rather, we infer the quark masses by various indirect means. For example, the measured mass of a free hadron gives a clue to the masses of the quarks it contains.

A quark and antiquark can be created from an energetic gluon in the interior of a hadron. Normally, when the quark and antiquark fly apart, other quarks and antiquarks are created and attach themselves to the original quark and antiquark, forming hadrons. Therefore, what we observe are not the free quark and antiquark but additional hadrons. The t quark, as we have said, is an exception, and the reason is that its lifetime is so short. After t quark and its antiquark are created, they rapidly decay into other particles before other quarks and antiquarks have time to attach themselves to the t and $\bar{t}$. So we infer the mass of the t from the energies of the particles it decays into.

The reason we do not know the masses of the neutrinos precisely is entirely different. Neutrinos can exist as free particles, but their masses are so tiny that, so far, they have been too small to measure. So we can say that neutrino masses are very small, much smaller than the mass of the electron, which has the smallest mass of the known charged particles.

In Table 16.2 we give the rest energies of the quarks and leptons, in the cases in which they are known. We do not list the rest energies of red, green, and blue quarks separately because the standard model says that quarks of a given flavor and different colors have identical masses. Furthermore, there is no experimental evidence that the mass of a quark of a definite flavor is different for different colors. So at present, we *assume* that all three colored quarks of a particular flavor have the same mass.

Table 16.2: Lepton and quark rest energies in MeV. The masses are the rest energies divided by c^2

Fermion	Rest energy (MeV)
e	0.511
μ	106
τ	1777
ν_e	$< 2 \times 10^{-6}$
ν_μ	$< 2 \times 10^{-6}$
ν_τ	$< 2 \times 10^{-6}$
u	3 ± 2
d	5 ± 3
s	100 ± 30
c	1250 ± 90
b	4200 ± 70
t	$174,000 \pm 3000$

In addition to the matter particles and the force carriers, the standard model requires the existence of still another particle, called the Higgs boson after the physicist Peter Higgs. (Other physicists also contributed to the idea.) So far, the Higgs boson has not been observed experimentally, but there is a good theoretical reason for believing it exits.

We now give some information about quantum field theory that will enable us to introduce the theoretical reason for the Higgs boson. As we have stated earlier, the standard model is a quantum field theory. It is a theory with many invariances (or symmetries), including translational invariance and rotational invariance. By translational invariance, we mean that the theory predicts that the results of an experiment will be the same even if the experimental apparatus is moved (the technical term is translated) to a different location. By rotational invariance, we mean that the the theory predicts that the results of an experiment will be the same even if the apparatus is rotated to a different orientation on the earth. These are natural invariances for a theory to have. In fact, science as we know it would not be possible if an experiment in New York and an experiment in London would give different results.

But there is another invariance in quantum field theory that is not as obvious as translational and rotational invariance. The other invariance is called "gauge invariance." As with so many names in physics, this one may be misleading, and so it needs explanation. As we have said, in quantum mechanics a wave function is in general a complex number, which means that it has a real and an imaginary part. A complex number that has absolute magnitude unity is called a "phase." An overal phase in a wave function is not observable because, as we have said, probability depends only on the absolute square of the wave function. In quantum field theory also, some of the fields are complex quantities. It stands to reason that the overall phase of a quantum field should not be observable. If it is not, then the field theory is invariant under a change in the phase of the field. Under such circumstances, instead of the theory being called "phase invariant," for historical reasons that don't concern us, the theory is called "gauge invariant." The standard model is a gauge invariant theory.

Now one fundamental aspect of a quantum field theory is that if it is to be gauge invariant, it must not only have matter fields (like the electron field) but also force fields, like the electromagnetic field. These force fields are called "gauge fields" because they are required by the gauge invariance of the theory. Other gauge fields in the standard model are the gluon fields of the strong interaction and the $W^{\pm 1}$ and Z^0 fields of the weak interaction. As we have seen, the quanta of the electromagnetic gauge field are the photons, which have zero mass. The gluons, which are the quanta of the strong field, also have zero mass. It is a general property of the gauge fields that their masses must be zero, because a non-zero mass in general destroys gauge invariance.

But the $W^{\pm 1}$ and Z^0 fields have masses that are not zero; in fact, they are large compared to the mass of a proton. How can that be? Peter Higgs and other physicists discovered that by introducing a new scalar field into the theory (called the Higgs field), they were able to maintain gauge invariance while giving the quanta $W^{\pm 1}$ and Z^0 masses. The gauge invariance (or gauge symmetry) is hidden and is called a "spontaneously broken symmetry." It was also discovered that if the quarks and leptons start out with zero masses, the same Higgs field can be put into the theory in a way that will give masses to them as well.

The Higgs bosons (the quanta of the Higgs field) also have a mass greater than zero. The mass of a Higgs boson is estimated to be more than 100 times as great as the mass of the proton, and so far the Higgs boson has not been discovered. A giant accelerator is being built in the CERN laboratory near Geneva, Switzerland. The accelerator is called the Large Hadron Collider or LHC, and in it a beam of high-energy protons will be made to collide with another beam of high-energy protons going in the opposite direction. This accelerator is scheduled to be finished some time in 2007, and should be ready for experiments in 2008. Many physicists hope that Higgs bosons will be discovered there. If not, the standard model will probably have to be modified in an important way.

In the next chapter we turn away from the very small and focus on the very large. We start with the sun and the solar system and in subsequent chapters build our way up to the universe as a whole. In understanding the very large, our knowledge of the very small is useful.

Chapter 17

The Sun and the Solar System

And yet it does move.

Galileo Galilei

After Galileo was tried for his heretical book in which he implied that the earth moved around the sun, he recanted, but is supposed to have muttered under his breath what he really believed about the earth, saying, "And yet it does move." Poor Galileo's recantation did not prevent him from being confined to house arrest for the rest of his life.

17.1 The sun

As is well known today, the sun is by far the largest object in the solar system. Its diameter is about 109 times as large as that of the earth, and its mass is about 333 thousand times as great as the mass of the earth. Its surface temperature is about 6000 K (10,300 degrees Fahrenheit) and the temperature of its core is about 10 million K.

We calculate the sun's diameter by measuring its distance and apparent diameter. We calculate its mass from Newton's law of gravity. We obtain its surface temperature by assuming the sun is a blackbody, so that the spectrum of light it emits depends only on its temperature.

To obtain the central temperature of the sun, we have to make a model of how it shines. The standard solar model assumes that the sun emits light because of a variety of nucler fusion reactions in its core. The rate of these fusion reactions depends strongly on the temperatures in the core, so that we can deduce these core temperatures by how much energy the sun gives off. The photons created by the fusion reactions in the core are not emitted right away at the surface. The photons are emitted, absorbed, and re-emitted randomly, and take perhaps a million years on average to get from the core to the surface. In Figure 17.1 we show a schematic picture of the interior of the sun, based on the model of how it shines. The radius of the core is about one-fourth the radius of the sun. The layer near the surface is called the "convective zone," and gas circulates between the inner and outer boundaries of that zone. Between the boundary of the core and the inner boundary of the convective zone, energy is transported mainly by electromagnetic radiation.

After reaching the surface of the sun, the photons take an additional eight minutes to go through empty space from the sun to the earth, a distance of about 93 million miles or 150 million kilometers (1.5×10^8 km). To be more specific, this distance is the mean length of the semimajor axis of the ellipse which the earth traverses in going around the sun. Astronomers call this distance the "astronomical unit," abbreviated AU. Approximately, the average distance between

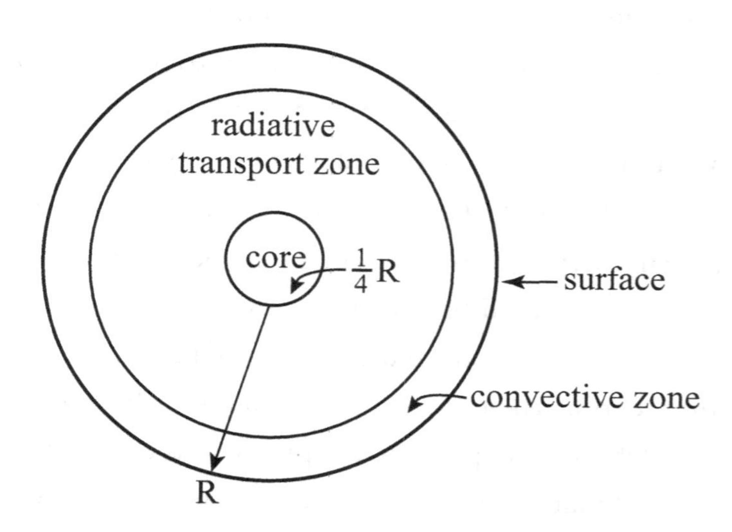

Figure 17.1: Schematic drawing of a cross section of the interior of the sun.

the sun and earth is 1 AU.

The sun remains the same size because the energy given off in the fusion reactions, which would tend to blow apart a small body, is balanced in the sun by the attraction of gravity. In addition to emitting electromagnetic radiation, mostly in the form of visible light, the sun emits electrically charged particles, called the "solar wind." These particles, consisting mostly of electrons and protons, keep moving outward to distances much farther than the earth. It is the earth's magnetic field that deflects these particles away from the surface of the earth.

In most of the fusion reactions in the sun, neutrinos are emitted. Unlike the photons, most of the neutrinos go directly from the core to the surface at almost the speed of light. For some years, it has been possible to detect neutrinos from the sun here on earth. Very large detectors have to be used because most neutrinos go through the detectors without interacting. In fact, most neutrinos incident on the earth go right through it without interacting. Yet so many neutrinos are emitted by the sun that some of them interact in our detectors.

A test of the standard solar model is that the number of neutrinos detected on earth should be calculable from the number of nuclear reactions necessary to produce the amount of light from the sun observed from earth. The early experiments found that too few neutrinos were detected compared to the predictions of the solar model. This fact meant that either the standard solar model is wrong or that something happens to the neutrinos on their way from the core of the sun to the earth. We now know that something happens to the neutrinos. The neutrinos emitted by the core of the sun are electron-type neutrinos ν_e because they are created with electrons (actually antielectrons, or positrons). The early detectors were designed in such a way that they could detect ν_e but not ν_μ or ν_τ. So if some of the neutrinos coming from the core of the sun convert into another type of neutrino before arriving on earth, the early detectors would not have been able to detect those neutrinos. We now have detectors that are sensitive to all three kinds of neutrinos, and have measured that the number of neutrinos reaching the earth is consistent with the prediction of the standard solar model. The changing of one kind of neutrino to another is called "neutrino oscillation."

We see that a knowledge of elementary particle interactions on the microscropic scale enables us to understand the processes that make the sun shine.

The sun and the planets are believed to have condensed from a giant gas cloud, consisting mostly of hydrogen gas but containing many other elements, especially helium. Near its surface, the sun is almost 80 per cent hydrogen and most of the rest helium, with small amounts of other gases and metals. The core of the sun must contain a smaller percentage of hydrogen and a larger percentage of helium because of the fusion reactions that have already taken place, converting hydrogen into helium. The outer layers of the sun are similar in composition to the composition of the giant planets Jupiter and Saturn, adding strength to the idea that the sun and planets formed from a single cloud of gas. The inner planets, like the earth, do not have much hydrogen and helium gases because gravity on these smaller planets is not sufficient to hold these gases. Thus, these elements have escaped from the earth and the other inner planets. Most of the hydrogen on earth is bound to oxygen as water. There is also a little helium in the interior of the earth.

The sun is believed to be 4.6 billion years old, the same age as the earth. That is because the model of how the solar system formed puts the time of formation as very short compare to the time of existence.

How do we know the age of the sun and earth? We know this because of meteorites, small rock or iron bodies that have fallen on earth from elsewhere in the solar system. These meteorites were presumably formed at the same time as the sun and the earth. Some meteorites contain radioactive elements with half-lives of several billion years. Measurements of the amount of these radioactive elements together with measurments of the amount of their stable decay products (which are right beside the radioactive materials) enable us to calculate the age of the meteorites quite accurately.

Evidence that the sun is a ball of gas comes from the fact that it rotates faster at the equator than at middle latitudes, and even slower at its poles. The time of rotation at the equator is 25 days. Astronomers can deduce the rotation time by observing how long it takes for sunspots to move across the face of the sun.

Near the surface of the sun are gases of elements much cooler than the surface temperature, and these absorb light. As we have previously noted, excited atoms emit light at certain definite frequencies, producing what are known as line spectra. The line spectrum of an element is characteristic of that element. Likewise, atoms absorb light at the same frequencies that they emit. Thus, the atoms absorb light, giving rise to dark lines in the otherwise continuous spectrum of the sun. These absorption spectra are also characteristic of the particular element. The element helium was given its name because it was first observed in the sun (helios is the Greek word for the sun) from helium's absorption line spectrum.

Many additional properties of the sun have been studied in detail, but we do not have the space here to discuss them.

17.2 The planets

For many years, astronomers knew of eight planets in the solar system. Starting from the planet nearest the sun and going outward, the planets are Mercury, Venus, earth, Mars, Jupiter, Saturn, Uranus, and Neptune. The four innermost planets are relatively small and rocky. The outer four planets are relatively large and probably contain mostly hydrogen and helium. The inner planets may have once contained hydrogen and helium, but because of their relatively small size, gravity is too weak to have retained those light gases in any but trace amounts.

In 1930 the situation changed with the discovery of a small planet-like object orbiting the sun beyond the orbit of Neptune. The object was named Pluto and called a ninth planet. The situation changed once again on August 24, 2006, at a meeting of the International Astronomers Union in Prague, in which Pluto was demoted from a regular planet to a "dwarf planet," a status that it now shares with several other planet-like objects. At the meeting, a planet of the solar system has to orbit the sun (rather than a planet), be large enough to be approximately spherical because of its own gravity, and be large enough to "dominate" its orbit, which means to clear its orbit of most of the debris of small objects orbiting the sun (except for satellites orbiting the planet itself). Later in this section we discuss the eight major planets and then discuss the dwarf planets, including Pluto. The name "dwarf planet" is ugly and may be changed. We prefer "minor planet" but shall use the name bestowed by the International Astronomers Union.

As we have already said, as stated in Kepler's first law, all the planets travel in ellipses (actually, approximate ellipses), with the distance of closest approach to the sun called the perihelion, and the farthest distance from the sun called the aphelion. The line between the perihelion and the aphelion is twice the semimajor axis of the orbit. All planets shine because of light reflected from the sun. The fraction of light reflected is called the "albedo."

The planets formed, like the sun, from the condensation under gravity of a cloud of gas and dust. The fact that the planets are very nearly spheres and the fact that their orbits are close to ellipses follow from the law of gravity. The planets do not fall into the sun because they have angular momentum, just like the moon does not fall onto the earth because of its angular momentum. Evidently, the original cloud of gas and dust was rotating, and in order to conserve the angular momentum of rotation, a small part of the gas and dust had to condense into planets rather than all of it condensing into the sun.

Astronomers have looked at the properties of many other stars in our galaxy, and these properties have provided a guide as to what will eventually happen to the sun. Most stars exist for billions of years in a relatively stable equilibrium, with outward pressure arising from nuclear fusion of hydrogen balancing the inward pressure arising from gravity. But eventually most of the hydrogen gets used up, and the energy produced by the star is mostly by the fusion of helium. It is not too long after the exhaustion of hydrogen in a star similar in size to the sun that it becomes a red giant, expanding to many times its original size. It has been estimated that after perhaps five more billion years, the sun will become a red giant, and its size will increase so much that it will envelop at least the innermost planet Mercury. At that time the earth will be little more than a large, spherical cinder.

Mercury is the planet nearest the sun. The deviation of the orbit of Mercury from an ellipse arises, as we have said, both from perturbations from the other planets and from corrections to Newtonian gravity from general relativity. The deviation is measured as a precession of the perihelion (position of closest approach to the sun) of Mercury as it goes around the sun, which means that the perihelion is in a slightly different position with each orbit. It takes Mercury about 88 days to complete one orbit around the sun, so Mercury's "year" is much shorter than the earth's year of 365 days. It takes Mercury about 59 days to rotate once on its axis, so Mercury's "day" is much longer than an earth day. As we have noted previously, the closer a planet is to the sun, the shorter its year must be, in accordance with Kepler's third law of planetary motion, which can be deduced from the law of gravity.

Mercury's diameter is about 3480 km, which is about 27 per cent as big as the earth and a little bigger than our moon. However, Mercury is denser than the moon, so that its mass is almost five times as great. The length of Mercury's semimajor axis is about 58 million km, or 0.39 AU. (Recall that 1 AU is the length of the earth's semimajor axis.) Because Mercury is so near the sun, it rises and sets within two hours of the sun.

Mercury has a magnetic field, a fact that must arise from an iron core. The field is only about 1 per cent as strong as the earth's magnetic field. Mercury does not reflect much light from the sun, its albedo being only 0.06. Mercury's day side is at a temperature of about 700 K, while it's night side is at a temperature of about 100 K. (Recall that the freezing point of water is 273 K, and the boiling point of water is 373 K). So Mercury is much hotter than boiling water during the day and much colder than freezing water at night. Two reasons for this large difference between the daytime and nighttime temperatures are the slow rotation of Mercury and its lack of an atmosphere.

Venus is the second planet from the sun. It is even hotter than Mercury, having a temperature of around 750 K day and night because of its thick atmosphere which holds in its heat in what is known as the greenhouse effect. Because of its cloud cover, its albedo is 0.76, the largest among the planets. Aside from the moon, Venus is the brightest object in the nighttime sky, somewhat brighter than Jupiter (because it is closer to the earth than Jupiter) and quite a bit brighter than the brightest star. The semimajor axis of its orbit is 0.72 AU, or about 108 million km. Because Venus is closer to the sun than the earth is, it is never too far from the sun when we view it. We therefore observe Venus either in the evening or in the early morning, and Venus is often called either "the evening star" or "the morning star," even though it is not a star at all.

Venus is similar to the earth in size and mass, being a little smaller than the earth in both. Its diameter is 95 per cent that of the earth, and its mass is 85 per cent of the earth's mass. Its surface gravity is 90 per cent as large as gravity at the surface of the earth. So if a person could stand the heat on Venus, he would weight 90 per cent as much as he would on earth.

One property of Venus that is quite different from that of the earth is its period of rotation, which is 243 earth days, and is retrograde (backwards, or in the opposite direction to the direction of its orbital motion around the sun). Venus's orbital period is only 225 days, so Venus has the peculiar property that its day is longer than its year.

(The day and year are here defined as the period of rotation about its axis and revolution about the sun respectively.)

Because Venus and Mercury are closer to the sun than the earth is, we do not always observe them as round. At times parts of them that would be visible from the earth if they were light, do not receive sunlight, and so are dark. So Venus and Mercury exhibit phases like our moon does, sometimes being full, other times being dark, and most of the time being part light and part dark and so looking like only part of a disk. When Venus is on the near side of the sun (with respect to the earth), it looks bigger (because it is closer to earth) than when it is on the far side of the sun, but on the near side part of Venus is in darkness. Galileo was the first person to observe the phases of Venus.

Mercury and Venus are the only two planets without any satellites. Most planets have more than one, but the earth, as we know, has only one: the moon. Of course, by a satellite, we mean a natural satellite; since the space age, a variable number of artificial satellites orbit the earth.

The earth, of course, is the planet that we know most about. One could write a whole book about the properties of earth, but here we concentrate on just a few of its features. Of all the planets, the earth has the greatest density, 5.52 grams/cm^3. (To get an idea of how dense this is, we note that water has a density of 1 gram/cm^3, so that the earth is more than five times as dense as water. Rock is more dense than water but less dense than the average density of the earth, and we believe that the earth's core is composed largely of liquid and solid iron. The iron at the earth's core is responsible for the earth's magnetic field.

The orbital period of the earth is of course one year, which is 365.26 days. Because the year is a little more than 365 days, once every four years the calendar has an extra day, which is February 29, and we call the year a leap year. Even that is not quite right, and other adjustments are very occasionally made.

The semimajor axis of the earth, by definition, is 1 AU (astronomical unit), which is about 150 million kilometers or 93 million miles. The earth's orbit, although eliptical, is very nearly circular (as are the orbits of most of the planets). The perihelion distance is 0.983 AU, while the aphelion distance is 1.017 AU. In the northern hemisphere, summer usually arrives on June 21, which is the longest day of the year. Actually, the earth is farther away from the sun on June 21 than on the first day of winter in the northern hemisphere, December 21. The reason that the days are longer in summer than in winter is that the earth's axis is tilted by 23.3 degrees from the ecliptic. In summer, the northern hemisphere tilts toward the sun, while in winter the southern hemisphere tilts toward the sun.

The earth rotates on its axis once every 23 hours and 56 minutes. Our 24-hour day is defined in relation to the position of the sun as seen from earth, say, from noon until noon. The day thus defined is a little longer than the period of rotation because as the earth is rotating it also is orbiting the sun, and that leads to the difference.

The escape velocity at the surface of the earth is 11.2 km/s (about 25,000 miles per hour). This fact means that in the absence of air resistance, an object going straight up from the surface of the earth at more than 11.2 km/s will never fall back to earth.

The earth's atmosphere consists of about 80 per cent nitrogen and 20 per cent oxygen, plus other gases, such as water vapor and carbon dioxide, in relatively small amounts. Although there is much less than 1 per cent of carbon dioxide in the atmosphere, this gas plays a crucial role on earth. Using energy from the sun, plants containing chlorophyll take up carbon dioxide from the atmosphere and combine it with water to produce carbohydrates (sugars, starches, and cellulose) in a process known as photosynthesis. In addition, carbon dioxide is a greenhouse gas, which traps energy from the sun, making the surface of the earth warmer than it would otherwise have been. For many centuries the amount of carbon dioxide in the atmosphere has varied between rather small limits. However, the burning of fossil fuels became large enough in the second half of the twentieth century and continuing into the twenty-first century to increase substantially the amount of carbon dioxide in the atmosphere. This man-made activity of burning coal and oil is a cause of global warming, which may have disastrous effects on the ecology of the earth if the burning of fossil fuels is not substantially reduced.

We know quite a lot about the earth's interior, mostly from the seismic signals from earthquakes. The interior of the earth is hot, partly because the earth was hot when it was first formed, and partly because radioactive materials in the earth's interior still generate heat when they decay.

According to geologic evidence, life has existed on earth for about three and a half billion years, so evidence for life appeared about one billion years after the earth was formed. Life itself may have appeared even earlier than the evidence shows. Models of the sun indicate that the sun is very slowly getting hotter, so that in about one billion years from now, the sun will be hot enough to evaporate the water on earth and extinguish life as we know it. If these models are right, life has already used up more than three fourths of the time of its existence.

Of course, life could become extinct much sooner than a billion years from now, possibly because of human activity, such as a nuclear war. It is also possible that a collision with an asteroid, a comet or some other large object from outer space could make life on earth extinct. In the history of the earth since life began, a number of objects from outer space have collided with the earth, some of them leaving craters that are still partially visible today despite erosion. Some of these objects were big enough to cause a large number of extinctions of species, but so far none of the objects was big enough to make all life extinct. We just have to hope that in the future, life will be as lucky as it has been in the past. We do not discuss still other possible scenarios for the extinction of life on earth.

Soon after the earth was formed and well before there was any life on earth, most astronomers believe that a planetary sized object collided with the earth. If there had been life at the time, it would have been extinguished. In the collision, a large amount of debris was lifted into orbit around the earth. It is believed that these pieces of debris coalesced because of gravity and became our moon. One of the reasons for this belief is that it is hard to envision a scenario in which the moon could have been captured by the earth. Another reason is that the composition of the surface of the moon is remarkably similar to that of the surface layers of the earth. The density of the moon is about 3.3 gm/cm^3, which is approximately the density of rock on earth. We conclude that the moon does not have much or any of an iron core.

The moon is about 384,400 km (about 240,000 miles) from the earth, or about 60 earth radii away. The radius of the moon is a little more than a quarter of the radius of the earth, and its mass only about 1.23 per cent of the mass of the earth. The surface gravity of the moon is only 1/6-th that at the surface of the earth, so an astronaught on the moon has only 1/6-th of his weight on earth. The moon is too light to have retained any atmosphere it may have once had, so it is essentially without any atmosphere at all. Because of the lack of atmosphere to moderate temperatures, the surface of the moon is very hot during the day and very cold at night. Another consequence of the lack of atmosphere is that there is no "weathering" of craters formed by collisions of meteors and comets with the moon. Therefore, the moon's surface is pockmarked by these craters.

Because of tidal friction, the moon's rotation has slowed over the years, so that now it always keeps the same face (called the near side) toward the earth. We never see (from the earth) the other side of the moon (called the far side). In addition to mountainous areas, the near side of the moon contains some relatively smooth, flat areas with not so many craters. These flat areas are called "maria," Latin for seas, because early astronomers thought they were bodies of water. Now we know that the moon has no liquid water on its surface.

During the years beginning in the 1950s, a number of space ships from the United States and the Soviet Union (a country subsequently split up, with the largest piece being the present nation of Russia) orbited the earth and explored the moon, including the far side, which is more mountainous than the near side. During the years from 1969 through 1972, the United States sent six manned space ships to the moon and back, bringing back not only information but samples of the lunar surface.

The moon is of course much smaller than the sun, the sun having about 400 times the radius of the moon. But the sun is much farther away from us than the moon is, and, by a peculiar coincidence, the sun and moon have the same angular size as seen from earth. This enables the moon to eclipse the sun on occasion. Sometimes the eclipse is total, with the entire disk of the sun behind the moon. Because the orbits of the sun and moon are ellipses rather than circles, sometimes the distances of the moon and sun are such that the angle subtended by the moon is slightly smaller than the angle of the sun. If there is an eclipse of the sun at that time, it is called an "annular eclipse," because the edge of the sun sticks out all around the moon, making a bright ring or annulus as seen from earth. There are also partial eclipses of the sun, in which the moon covers only a portion of the sun's disk.

Mars is the fourth planet from the sun. It is slightly reddish to the eye, and has been named "the red planet." Mars has a thin atmosphere and dust storms. As far as we know, there is no liquid water on the surface of Mars. Unmanned landings on Mars have revealed a barren surface with no signs of life. Because only a small part of Mars has been explored, there isn't any proof that there is no life on the surface or under the surface of the planet, but from what we know, it seems unlikely.

The semimajor axis of the orbit of Mars is about 1.5 AU (one and a half times that of the earth). The Martian year is 687 days (almost 1.9 earth years). The Martian day is about 24 and a half hours, very similar to the earth day, and the tilt of the Martian axis of rotation is very similar to that of the earth. Because of this tilt, Mars has seasons similar to those on earth, but Mars, being farther from the sun, is much colder than the earth, with temperatures varying from about -220° F to about 60° F. Near the equator at midday, the temperature on Mars is fairly comfortable, but its atmosphere cannot support the life of any creature that breathes air on earth.

Mars has only a little over half the diameter of the earth, and its density is closer to that of the moon to that of the earth. The mass of Mars is only about one-tenth that of the earth. Its surface gravity is a little less than 40 per cent of the surface gravity of the earth. Its albedo is 0.15.

The very thin atmosphere of Mars is mostly carbon dioxide. The planet has polar "ice caps," but these are mostly frozen carbon dioxide with relatively little water ice. Mars often has high winds and is subject to dust storms despite its thin atmosphere. It also has high mountains, the highest of which is about three times as high as Mount Everest, the highest mountain on earth. Of course, one reason that Mars has such high mountains is that its surface gravity is considerably smaller than earth's. That is not the only reason, because Mercury, which is smaller than Mars, does not have such high mountains. Other factors, not meant to be inclusive, are the makeup of the interior of the planet and its volcanic activity.

Mars has two satellites, Phobos and Deimos, both of which are very much smaller than our moon and considerably closer to Mars than the moon is to the earth. Both are irregularly shaped, being too small for gravity to have made them spherical. They are both elongated and each keeps one end toward Mars. Phobos, the larger moon, is only 28 km long and 20 km wide and has a period of less than eight hours. Deimos is farther away from Mars and has a period of somewhat more than 30 hours.

Jupiter is the largest planet in the solar system. It is the nearest to the sun of the four large gaseous planets, the others being Saturn, Uranus, and Neptune. Jupiter's diameter is more than ten times the diameter of the earth, so if Jupiter had the same density of the earth, it would have a mass of more than 1000 times as great as the earth. However, Jupiter's density is only 1.33 g/cm^3, much less than the density of the earth, which is 5.5 g/cm^3, so that Jupiter has a mass that is only 318 times as large as the earth's. Because of Jupiter's low density, astronomers believe that much of it is a giant gas ball, composed mainly of hydrogen and helium. Jupiter, despite its large size, is too small to have temperatures in its core high enough to cause nuclear fusion reactions. Therefore, like the other planets, it shines because of reflected sunlight. Its albedo has the rather large value of 0.51.

The semimajor axis of Jupiter's orbit is 5.2 AU. At first this distance was a puzzle to the early astronomers. The reason for the puzzle was that Jupiter's distance from the sun did not agree with the prediction of a "law" discovered in the latter half of the eighteenth century. This law, known as Bode's law (although Bode only popularized it rather than discovered it), gives the approximate distances of most of the planets from the sun. According to Bode's law, there should be a planet between Mars and Jupiter, but such a planet was not known. Some years later a small, planet-like object, was discovered at the approximate distance predicted by Bode's law. The new object was named Ceres, and was at first considered a new planet.

Subsequently many more small objects were discovered at approximately the same distance from the sun as Ceres, and we believe that there exist as many as 100,000 of them of various sizes. They are commonly called asteroids. Perhaps the nearness of Jupiter to these objects prevented them from coalescing under gravity to form one planet. Most of the asteroids are so tiny that they are irregular in shape, but Ceres (with a diameter of perhaps one twelfth that of the earth) is large enough to have become approximately spherical because of its own gravity. After other asteroids were discovered, Ceres was demoted from a planet to just another asteroid, but now it is considered to be a dwarf planet.

There is no known physical reason why Bode's law should be approximately true for most of the planets, and most scientists regard the planets as arising from the accidental conditions at the time the solar system was formed.

We return to Jupiter. Its period of rotation is under ten hours, so that the speed of movement of a point on its equator is much larger than that on earth. The rate of rotation is a little larger at the equator than at the poles. Because of its rapid rotation, Jupiter bulges out at the equator, its equatorial diameter being six per cent larger than its polar diameter. On the other hand, Jupiter's period of revolution around the sun is much longer than an earth year—in fact, it is a little less than 12 years, and agrees well with Kepler's third law, which relates the planet's distance from the sun to its period of revolution.

A view of Jupiter in a telescope reveals dark bands, which are features of its atmosphere. It also has on its surface a large oval-shaped reddish spot that is believed to be a long-lasting storm. The area of the red spot is larger than the size of the earth. What we observe as the surface of Jupiter is really its atmosphere. It is believed that most of Jupiter is fluid (mostly liquid) and that any heavy metals on the planet have sunk to its core. Below ordinary liquid hydrogen, hydrogen probably exists as a liquid metal owing to the very high pressure in the interior. Because we have not probed into the interior of Jupiter, our understanding of its interior has come from theoretical calculations.

The surface gravity on Jupiter is 2.64 times as great as on earth. If a human being could stand on Jupiter, she would weigh 2.64 times as much as she does on earth. However, an object placed on Jupiter would sink deep into it gassy atmosphere until it would presumably reach liquid.

According to recent accounts, Jupiter has at least 60 satellites, the four largest of which are comparable to or larger than our moon. These are the four that Galileo first discovered with his telescope. These moons are Ganymede (the largest, which is somewhat larger than the planet Mercury), Callisto (about the size of Mercury), Io (larger than our moon), and Europa (a little smaller than our moon).

Most of the other satellites are too small to be spherical and are irregular in shape. Because of tidal forces, all Jupiter's moons are thought to keep the same side toward Jupiter, just as our moon always keeps the same side toward earth. Of all Jupiter's satellites, Io is the closest to the planet, except for a few tiny moons. Because of its closeness to Jupiter, tidal forces cause volcanic activity on Io, observed by a spacecraft called Voyager, which passed not too far from Jupiter and its moons, and radioed its photos back to earth. In addition to its satellites, Jupiter has a thin system of rings circling the planet at the equator. These rings consist of thousands of small objects, some of them tiny, that orbit Jupiter individually.

Saturn, the second largest planet in the solar system, is the next one out from Jupiter. It is, on the average, almost twice as far from the sun as Jupiter is and almost ten times as far from the sun as the earth is, Saturn's semimajor axis being 9.54 AU. It takes Saturn over 29 years to go once around the sun.

The radius of Saturn is almost nine times the radius of earth, and more than 80 per cent of the radius of Jupiter. Yet the mass of Saturn is only 30 per cent of the mass of Jupiter because Saturn has the lowest density of all the planets, being only 0.68 g/cm^3 (less than the density of water). Because of its low density and large radius, the surface gravity is similar to that on earth, but, because of the gaseous nature of the planet, a solid object would sink well beneath the surface until encountering liquid. Saturn, like Jupiter, rotates rapidly, about once every ten and a half hours. Therefore, like Jupiter, Saturn bulges out at the equator. The albedo of Saturn, at 0.50, is very similar to that of Jupiter.

Saturn is distinguished from the other planets by the large rings that go around it at the equator. These rings are far more visible than the thin rings of Jupiter. The inner edge of Saturn's rings begins at a distance out from the surface of about 40 per cent of the planet's radius. The outer edge is about another 80 per cent more distant. Although the rings are very wide, they are very thin, and when Saturn presents itself to the earth in such a way that the rings are viewed edge on, the rings can barely be seen through a good telescope. There is a large gap and many much smaller gaps in the ring system. The rings consist of mostly of tiny pieces of debris that individually orbit Saturn, although some of the pieces may be more than a meter in size. The rings are very beautiful when viewed through a sufficiently large telescope. In Figure 17.2 we show a drawing of Saturn with its rings.

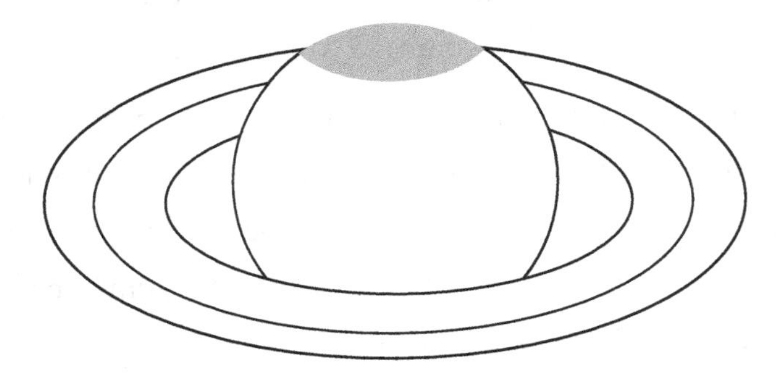

Figure 17.2: Drawing of Saturn with its rings. In a photo, Saturn and the rings would be bright against the dark background of the night sky.

Saturn has a large number of moons; well over 40 have been discovered so far. The largest is Titan, first discovered in the 17th century. Titan is just a little smaller than Ganymede, Jupiter's largest satellite. Titan is one of the few satellites in the solar system known to have an atmosphere. In fact, the atmosphere of Titan is quite thick, thicker than the earth's atmosphere. Several other satellites of Saturn are also large enough to be approximately spherical, but they all are considerably smaller than Titan. Of course, the many thousands of objects in Saturn's rings are also satellites of Saturn but are not normally called satellites.

Uranus, the next planet, was discovered accidentally in 1781 by the English astronomer William Herschel. (The six innermost planets were known well before the time of Galileo.) Uranus is classified as a gas giant, although it is considerably smaller than Saturn, having a radius equal to about four earth radii. Like the other gas giants, Uranus has a low density, just a little larger than the density of water. The semimajor axis of its orbit is a little over 19 AU, and its year is 84 earth years.

The period of rotation of Uranus about its axis is a little over 16 hours. Its axis of rotation is close to the plane of its orbit, so that the plane of the equator is almost perpendicular to the orbital plane. At the beginning of summer the north pole points almost directly at the sun and in the beginning of winter the north pole points almost directly away from the sun. Thus, the seasons are extreme on Uranus. It is not understood why the axis of rotation is so different from those of the other planets.

Uranus has at least 27 satellites, a few of which are approximately spherical. In addition, Uranus has several rings, which are much thinner and more tenuous than the rings of Saturn. The rings contain thousands of small objects.

Neptune was discovered in 1846, and not by accident. An examination of the orbit of Uranus over some years revealed deviations from an ellipse. If Newton's laws were correct, another planet, outside the orbit of Uranus, would have to be responsible. It was predicted independently by an Englishman, John C. Adams, and a Frenchman, Urban Leverrier, that the planet was in a certain small region of the sky. Astronomers looked there and found Neptune.

The semimajor axis of Neptune is 30 AU, and its year is 165 earth years. Neptune is a little smaller in radius than Uranus but somewhat more dense, so that its mass is a little larger than that of Uranus.

Neptune has 13 known satellites, the largest of which is Triton, which has a radius somewhat larger than half of our moon. The orbit of Triton is about six days. It is backwards, or retrograde (in the opposite direction that Neptune revolves around the sun). This retrograde motion makes it unlikely that Triton was condensed from debris near Neptune. The more probable explanation is that Triton was captured from an independent orbit around the sun. For such a capture to have occurred, Triton must have been near a third body (perhaps as part of a double system), which would have received energy and thrown into another orbit around the sun. Neptune also has a very thin system of rings, so that all the large gaseous planets have rings. However, only Saturn has a large and beautiful system of rings.

A dwarf planet is large enough so that gravity makes it approximately spherical but small enough so that it doesn't dominate its orbit by capturing most of the debris nearby. All dwarf planets are considerably smaller than Mercury, the smallest of the major planets. An example of a dwarf planet is Ceres, which coexists with many thousands of asteroids in the asteroid belt between the orbits of Mars and Jupiter. Ceres contains about one-third of the mass of the asteroids in the belt.

Another dwarf planet, Pluto, is in the innermost part of the so-called Kuiper Belt, named after Gerard Kuiper, who predicted in 1952 that bodies exist beyond Neptune. Objects in the Kuiper Belt, other than Pluto, were not discovered until the 1990s. There are many objects in the Kuiper Belt, and they are mostly objects of ice with some rock. Some of the Kuiper Belt objects are not far from Pluto's orbit, so that Pluto does not dominate its orbit and consequently fails to be a planet in the new scheme.

In 2005 Mike Brown, an American astronomer, discovered an object in the Kuiper Belt that is a little larger than Pluto. The object was officially named Eris in September, 2006. Like Pluto, Eris qualifies as a dwarf planet. There may be, in addition to Pluto and Eris, dozens of other Kuiper Belt objects, some of them still undiscovered, that would qualify to be called dwarf planets. As of the first half of 2006, Brown had already observed eight apparently round Kuiper Belt objects that might qualify, and some astronomers have conjectured that there are many more. There are many other objects in the Kuiper belt too small to be approximately spherical, and therefore don't qualify as dwarf planets but are just objects orbiting the sun.

Pluto has a large satellite called Charon plus two tiny satellites. For a technical reason, Charon may itself be classified as a dwarf planet. The reason takes some explaining. We begin this explanation by considering our own moon. It is usually said that the moon orbits the earth in an ellipse, but that is not strictly speaking true. Actually, both the earth and the moon orbit a point that is the center of mass of the earth and moon.

We now explain roughly what the center of mass is. To do so, consider a hypothetical case of two stars of equal mass that are under the influence of each other's gravity. It does not make sense to say that one star orbits the other in an ellipse, because which one would be at the focus of the ellipse and which one would do the orbiting? Actually, both stars orbit a point in space half way between the two stars. This point is called the center of mass of the two stars.

If one of the stars has a greater mass than the other, the center of mass is closer to more massive star. As the mass difference gets larger, the center of mass moves ever closer to the heavier star. Finally, if the difference is very great, the center of mass can be inside the heavier star, but not at its center. The lighter star will then orbit the center of mass in an ellipse, and the heavier star will just wobble a bit, orbiting a point inside itself. So it is with the earth and the moon. The center of mass of the earth and the moon is actually inside the earth, so the moon can be said to orbit the earth in an approximate ellipse.

The sun is so much heavier than the earth that the the center of mass of the sun and the earth is well within the body of the sun. However, the center of mass of the sun and Jupiter, the largest planet, is a little outside the surface of the sun on an imaginary line between the center of the sun and the center of Jupiter. For this reason, Jupiter is not like the other planets, but rather the sun and Jupiter are a binary system.

We return to Pluto and Charon. The masses of these bodies are such that their center of mass is in space outside of Pluto. Therefore, both Charon and Pluto orbit a point in space rather than Charon orbiting Pluto. In that situation, both Charon and Pluto can be considered dwarf planets orbiting the sun as they orbit around their center of mass. Neither Pluto nor Charon orbits the sun in an ellipse. Rather, it is the center of mass of Pluto and Charon that orbits the sun in an ellipse. However, we still regard Charon as a satellite of Pluto.

Chapter 18

The Milky Way

> One thing I have learned in a long life: that all our science, measured against reality, is primitive and childlike—and yet it is the most precious thing we have.
>
> Albert Einstein

As we have already said, the Milky Way is a galaxy containing more than a hundred billion stars. These are bound in the galaxy by their mutual gravitational attraction as well as by the gravitational pull of so-called "dark matter" in the galaxy. We first discuss the stars and other bright objects in our galaxy and then the dark matter.

18.1 The stars

A star is formed when a huge cloud of dust, consisting of mostly hydrogen and helium and a lesser amount of other elements, contracts under its own gravity. As the gas contracts, gravitational potential energy is converted into kinetic energy of the molecules of the gas. In other words, as the gas molecules fall, they speed up. Because the temperature of a gas is proportional to the average kinetic energy of its molecules, the gas gets hotter and hotter as it continues to contract. Eventually, the gas becomes so hot that collisions remove the electrons from the gas atoms, and the gas becomes ionized. Still later, the core becomes so hot that nuclear fusion reactions take place, leading to an equilibrium between the outward pressure from heat generated by the nuclear reactions and the inward pull of gravity. The star then settles down and becomes what is called a "main sequence" star.

The luminosity of a star is the total amount of electromagnetic energy it emits from its surface per second, including energy emitted at wavelengths not visible to the human eye. A main sequence star has a relationship between its luminosity and its surface temperature. The surface temperature can be deduced by the color of the star—the higher the temperature the bluer the star, and the cooler the temperature the redder the star. Stars in the main sequence vary in luminosity by huge amounts. Some supermassive stars are more than 100,000 times more luminous than the sun, while others are only 1 per cent as luminous. Most stars of the main sequence are in between these extremes. The most luminous stars have a surface temperature of up to 50,000 K (kelvins) compared to the sun's 6000 K, while the least luminous have surface temperatures of less than 3000 K (50,000 kelvins is about 90,000 degrees Fahrenheit).

A star that is bright to the eye was called by early Greeks a star of the "first magnitude." A star not so bright was called a "second magnitude" star, and so forth. More recently, astronomer have quantified the idea of magnitude, but the general idea remains. Because of this definition, the brighter an object in the sky, the smaller is its observed, or apparent, magnitude. Observed magnitudes of very bright objects, such as the moon and the very brightest stars and planets, are negative. According to the astronomers' definition of magnitude, a first-magnitude star is about 2.51 times as bright as a second-magnitude star, a second-magnitude star is 2.51 times as bright as a third-magnitude star, and so on. The sun, the brightest object in the daytime sky, has a magnitude of about −26. The brightest star Sirius has a magnitude of −1.4. Stars with magnitudes greater than about 5 cannot be seen with the naked eye, and so must be observed with telescopes.

Consider light emitted by a star impinginging on an imaginary spherical surface a certain distance away. The center of the sphere is at the position of the star. Now consider the light impinging on another imaginary spherical surface twice as far away. The farther surface has four times the area of the nearer one, so the intensity of the light (the amount of light per unit area) is only 1/4 as strong at the farther surface. Therefore, if we on the earth observe two stars, each with the same luminosity, but one twice as far away as the other, the observed brightness of the farther star is only 1/4 as great as the observed brightness of the nearer star. This fact illustrates the so-called inverse square law of observed brightness. Because of the inverse square law, if one star is three times as far away from us as another of equal luminosity, we observe the farther star as only 1/9 as bright as the nearer one, and so on.

The absolute magnitude of a star is the apparent magnitude corrected (using the inverse square law) to a standard distance of ten "parsecs." A parsec is a unit of distance commonly used by astronomers, and is equal to 3.26 light-years. (Recall that a light-year is the distance light can travel in one year.)

The luminosity of a star can be deduced by how bright it is to our eyes, or to our telescopes for the dimmer stars, and by how far away it is. We can tell the distance to nearby stars by parallax, that is, by seeing how much the star seems to shift its position when viewed on the earth at various positions in its orbit around the sun. (A more detailed discussion of parallax is given in Chapter 1.

For stars that are far away, the amount of parallax is too small to be detected, and other measures must be used. One such measure makes use of stars whose luminosity varies periodically. Such stars with periods of one to one hundred days, are called Cepheids. Observing nearby Cepheids, whose distance can be measured by parallax, we have learned that their luminosity depends on their periods. Therefore, by looking at a far-off Cepheid, we can tell its luminosity just by measuring its period. This measurement gives us the absolute magnitude of the star, and direct observation gives us the apparent magnitude. Knowing both, we can use the inverse square law to calculate the distance. Because Cepheids are bright stars, this method lets astronomers measure distances much farther away than use of parallax alone. An object that serves as a measurement of distance because its intrinsic brightness is known is called a "standard candle."

Another way to determine the distance of a star is to look at its position on the main sequence. The more luminous a star is, the hotter its surface temperature is. But the color of a star is related to its surface temperature: hotter stars are bluer and cooler stars are redder. Therefore, the absolute magnitude can be estimated from the color. If a bluish star is very faint (large observed magnitude), it must be very far away. Knowing both the absolute and observed magnitudes enables astronomers to calculate the distance.

A relatively small number of stars are not on the main sequence. Some are called "giants," often "red giants," that are very luminous but relatively cool. The only way that high luminosity can occur with cool temperature is if the star is very large—hence the name giant. The coolness is indicated by the reddish appearance of the star. Some cool stars have even greater luminosity than the giants, and are called "supergiants."

Still other kinds of stars are not on the main sequence. Among them are stars called "white dwarfs." These are stars that have relatively hot surfaces despite small luminosities. The only way such a situation can occur is by the star being smaller than usual—hence the name white dwarf.

Perhaps as many as half the stars in the galaxy are "binaries" or "double stars." A binary consists two stars orbiting around a common center of mass. Presumably, in many cases, stars have planets orbiting them. So far, astronomers have only observed a small number of extra-solar planets, or, in other words, planets orbiting stars other than the sun. It is very difficult to observe an extra-solar planet, as it is very small and it shines only by reflected light from its star. If a planet is large enough, it can cause a star to wobble, as the star and planet orbit their common center of mass. Such wobbly stars have been observed, and usually the planet causing the wobble is as large or larger than Jupiter.

18.2 Steller evolution

A star remains on the main sequence for many millions or billions of years until most of the hydrogen in its core is burned into helium. Paradoxically, the more massive a star is, the shorter the time that it exists on the main sequence. The reason is that the more massive stars are hotter and therefore burn their hydrogen more rapidly than the less massive ones, and so exhaust their fuel more quickly.

We believe that the sun is the same age as the earth because theory says that they were created at about the same time. If the theory is correct, then the sun is about 4.6 billion years old. (Radioactive dating using uranium decay tells us that the earth is 4.6 billion years old.)

We can predict that the sun will shine as a main sequence star for about 5 billion more years, based on the following reasoning. Models of the sun's interior, which do a good job of predicting the sun's surface temperature of around 6000 K, tell us that only the inner core, comprising about 10 per cent of the sun's mass, is hot enough to support fusion reactions. We also know that in the fusion taking place in the sun, about 0.7 per cent of the mass is converted into energy. We also can measure how much radiant energy the sun emits per second, so we know the rate at which the sun converts mass to energy. Knowing how much mass is available in the sun's core and the rate at which energy is produced, we can calculate how long the sun will be able to convert hydrogen to helium. The calculation yields about

10 billion years. Because the sun is a little less than 5 billion years old, we estimate that it will continue to shine as a main-sequence star for a little more than 5 billion additional years.

Using similar reasoning, we can calculate that a very massive star, with a much higher surface temperature, converts its mass to energy at a much higher rate, and may exist on the main sequence for only a few million years. On the other hand, a star much less massive than the sun will live much longer than 10 billion years. We shall see in the next chapter that the universe is about 14 billion years old. Sufficiently small stars will live much longer than the present age of the universe. All such small stars ever created are still on the main sequence.

Because very massive stars live such a short time (on the time scale of the universe), they are expected to be relatively rare. Likewise, because stars with very small masses live a long time, they are expected to be very abundant. Observation shows these expectations to be correct. In addition, the processes that create stars from dust clouds probably favors the formation of small stars. As there are still many dust clouds in the universe, new stars are still being created at the present time.

As a star with about the mass of the sun burns the hydrogen in its core into helium, the core gradually shrinks as the hydrogen is depleted because the core temperature is not high enough to fuse the helium into still heavier elements. As the core shrinks, it gets hotter because of released gravitational kinetic energy. (A rather poor analogy is that an object dropped from a tall building on earth acquires more and more kinetic energy as it falls.) Eventually all the hydrogen in the core is exhausted, but by this time the core is hot enough that a shell of hydrogen outside the core begins to burn. As the heating continues, the burning shell heats the region near the surface and forces it to expand and thereby cool. The expansion increases the radius of the star by perhaps around 100 times, causing the star to become a red giant.

When the sun becomes a red giant in about 5 billion years from now, it will expand enough to engulf Mercury and possibly Venus as well. The sun will probably not expand enough to engulf the earth, but all the earth's surface water will boil off before the sun reaches red giant size. We have already noted that models suggest that the sun will gradually get hotter while still a main sequence star, and that it is estimated that all life on earth will be extinguished in perhaps one billion years, well short of the 5 billion years the sun will remain on the main sequence. This slow warming of the earth over a period of a billion years is to be distinguished from the fast global warming over a time of perhaps 100 years caused by the increase of carbon dioxide caused by mankind's use of fossil fuels for energy. Short-term global warming should not be nearly sufficient to extinguish life on earth, but should be enough to make things difficult for mankind and to cause the extinction of many species of plants and animals.

Let us return to the evolution of a star like the sun. As the outer layers of the star expand, the core continues to shrink and heat up. The gas in the core contains many electrons, and these electrons cannot indefinitely be squeezed closer together in a star with about the mass of the sun. The reason is the Pauli exclusion principle, which, as we have seen, says that no two electrons can be in the identical quantum state. Eventually, the core is prevented from shrinking further by the pressure of the electrons resisting being squeezed into states forbidden by the Pauli princple. At this point, the gas in the core is said to be "degenerate." In a degenerate gas, the temperature can go up or down without the pressure changing. (In a sufficiently massive star, gravity will win out by forcing the electrons to combine with protons to form neutrons, thereby evading the Pauli principle.)

As the gas in the shell surrounding the core burns, it heats up the core until the helium fuses rapidly in what is called a "helium flash." The flash destroys the degeneracy of the core by exciting electrons to other states, and the star contracts and becomes hotter. When the helium in the core is exhausted, helium is burned in a shell around the core, and the star expands and cools again, becoming a red giant a second time. The core becomes degenerate a second time, and the giant gradually blows off its outer layers. What is left is a small star called a white dwarf. The white dwarf is not hot enough to fuse carbon and heavier elements, and its hydrogen and helium are largely used up, so it slowly cools while giving off electromagnetic radiation.

A white dwarf cannot exist with a mass greater than 1.4 times the solar mass because in a more massive star gravity is too strong for electron degeneracy to prevent the star from collapsing further. If a massive star does not eject enough material to keep its core below 1.4 solar masses, the collapse may lead to a huge explosion called a "supernova." The huge amount of heat generated creates heavy elements. Paradoxically, in the core the heat breaks up some of the heavier nuclei into protons and neutrons. The electrons are forced so close to the protons that the weak interaction may cause the electrons to combine with the protons, resulting in neutrons plus neutrinos. The neutrinos escape from the star, while the neutrons may form a stable degenerate state. (Neutrons also obey the Pauli principle.)

If these events take place, the result is a neutron star. A neutron star has a huge density, having a mass of more than 1.4 solar masses and a diameter of perhaps 10 km. The density is such that a cubic centimeter has a mass of about a billion tons.

In a still more massive star, the collapse probably always leads to a supernova. If the remains are above 1.4 solar masses and below about 2 solar masses, a neutron star results. However, if the remains are well above 2 solar masses, gravity is too strong for the neutrons to form a stable system, and the star continues to collapse to form a black hole, discussed in Chapter 10. We here reiterate that gravity is a consequence of curvature of spacetime, so that for sufficiently massive objects not even the Pauli principle can prevent a collapse. The star may become so hot that other forms of matter are created and many of the neutrons may be excited to other states. What ultimately happens to matter inside the horizon of a black hole is unknown.

18.3 The galaxy

As we have said, the Milky Way is a galaxy containing over a hundred billion stars. It can be seen across the night sky with the naked eye on a clear night. Studies with telescopes show that our galaxy is shaped somewhat like a disk, with a central bulge and spiral arms. In the middle of the central bulge is a dense core, believed to be a black hole with a mass more than a million times that of the sun. Matter keeps falling into the black hole, increasing its mass. As charged particles fall toward the black hole, they accelerate, giving off electromagnetic radiation in the form of X-rays, which can be detected on the earth. These X-rays, coming from an extremely small region in space, constitute evidence for the black hole.

Although most of the stars of the galaxy are in the disk, some are above and below it, in a kind of halo. In the halo are clusters of stars, called "globular clusters," containing up to hundreds of thousands of stars in roughly spherical regions.

Our galaxy is around 100,000 light-years in diameter, and our sun is in one of the spiral arms, around 30,000 light-years from the galactic center. In Figure 18.1 we show an artist's drawing of how our galaxy the Milky Way might look to an observer looking at it from the outside. The drawing is only illustrative and does not correspond in detail to what we know about the Milky Way.

The sun and most of the stars in the galaxy are rotating around the center. The Doppler effect may be used to measure the speeds of rotation. These measurements have led to the discovery that our galaxy contains "dark matter" in addition to the luminous matter we see.

The argument for dark matter goes as follows: According to Newton's law of gravity (general relativity is not necessary here), the speed of an object rotating around a massive body decreases as the distance to the massive body increases. Thus, for example, the earth revolves around the sun more slowly than Venus does, and Mars revolves more slowly than the earth does. Returning to the galaxy, we should expect stars far from the center of the galaxy (where most of

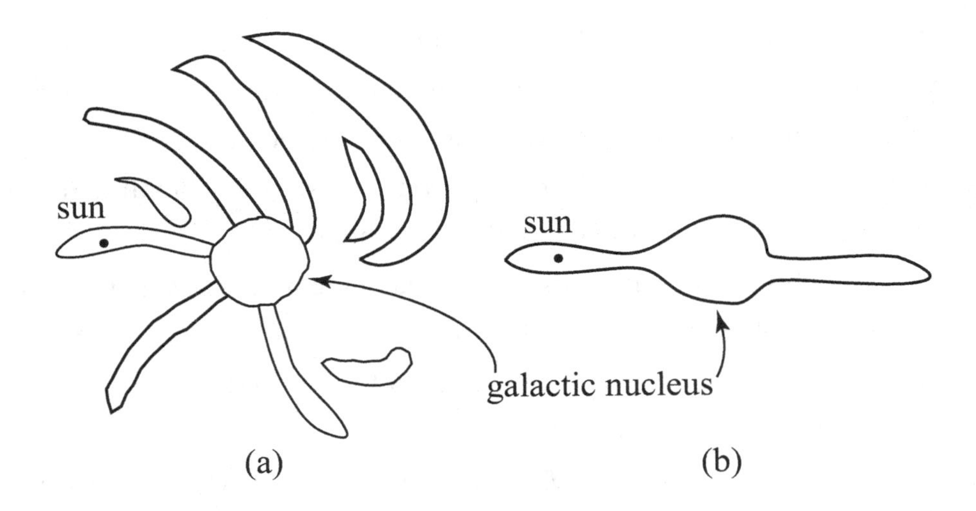

Figure 18.1: Drawing of the Milky Way as it might be observed from outside. (a) looking down at the plane of the galaxy. (b) looking edgewise at the galactic disk.

the luminous mass is located) to have smaller rotational speeds as their distance increases. We find, however, that the stars farther from the center rotate almost as fast as the nearer ones. This fact means that there is a large amount of additional unobserved mass, extending outward beyond most of the mass of the visible galaxy. Most of the unobserved mass is dark matter. We shall discuss dark matter in more detail in the next chapter.

Chapter 19

The Universe

> When the heavens were a little blue arch, stuck with stars, methought the universe was too straight and close: I was almost stifled for want of air: but now it is enlarged in height and breadth, and a thousand vortices taken in. I begin to breathe with more freedom, and I think the universe is incomparably more magnificent than it was before.
>
> —Bernard le Bovier de Fontenelle (1657–1757)

19.1 Expansion of the universe

If we look up into the sky with a powerful telescope, in addition to stars we see many fuzzy luminous regions. Some of these regions are luminous clouds of hot gases in our own galaxy, but others are galaxies something like our own, too far away for us to distinguish the individual stars. It is now estimated that there are hundreds of billions of individual galaxies in the visible universe. Some of these are spiral shaped much like our own galaxy, some are elliptical in shape, and still others are irregular. An elliptical galaxy is shaped something like an egg, although not necessarily smaller at one end than at the other, and more nearly spherical than an egg.

In the early twentieth century, the universe as a whole was believed to be static on the average, although it was known from the Doppler effect that individual stars in our own galaxy moved. When Einstein applied his general theory of relativity to the universe as a whole, he found to his dismay that the equations permitted a universe that was either expanding or contracting, but could not accomodate a static universe. In order to make his theory compatible with a static universe, Einstein modidfied his theory by adding a constant term, called the "cosmological constant," that could be adjusted to counteract the expansion or contraction and make the universe static.

It was not many years after Einstein introduced the cosmological constant that the American astronomer Edwin Hubble (1889–1953) discovered that the light from distant galaxies was redshifted. This fact implied that the far-off galaxies were not only moving but that their motion was not random—they were all receding from us. Furthermore, their speed was approximately proportional to their distance away from us. These facts could be explained if all the material we observe was concentrated at one point (or a very small region) in space and then at one time suddenly exploded. If so, the faster the particles moved, the farther away they would be. The astronomer Fred Hoyle, who did not believe that the universe began in an explosion, called it mockingly, "the big bang," and the name has stuck. When Einstein heard that the universe was expanding he said his introduction of the cosmological constant was his biggest mistake.

However, he was premature in discarding the constant, as later developments have shown. We shall discuss these developments later in this chapter.

We have to discuss several issues in connection with the big bang. First, if the distant galaxies are all receding from the earth, does that make the earth the center of the universe? The answer is no—the galaxies are all receding from one another. We can get help in understanding this fact by means of an analogy. Consider a balloon with dots marked on its surface. Then blow up the balloon. As it expands, every dot gets farther away from every other dot, so no dot can be considered as the center of the expansion of the balloon. The expansion of the universe seems to be similar, except that it is three-dimensional space that is expanding rather than a two-dimensional surface as on the balloon. We conclude that the earth and the sun are not in any special place in the universe, just as they are not in any special place in the Milky Way.

A second issue concerns how we know that the farther away a galaxy is, the faster it is going. We obtain the speed from the amount of the redshift, but we need a way to measure the distance. For objects such as far-off galaxies, individual stars, even Cepheids, are too faint to be observed. We have to look at extremely bright objects whose distance can be measured. These are supernovae.

There are several kinds of supernovae. The explosions of massive stars, which we have discussed earlier, are called type 2 supernovae. These are unsuitable for measurement of distance because their intrinsic brightness varies from one to another. We need to find bright objects, all of which have the same intrinsic brightness, so that we can calculate the distance of each one from its apparent brightness. Another kind of supernova, called type 1a, fills the bill. A supernovae at its brightest may be as luminous as the entire rest of the galaxy with its many billions of stars. Typically, a type 1a supernovae retains its brightness for a few weeks.

A type 1a supernova arises as follows: Consider a binary star system, one of which is a white dwarf and the other is a much larger star. In some cases, the gravity of the white dwarf is sufficiently strong to pull off the outer layers of the larger star and attract these layers to the surface of the dwarf. As this process continues, the dwarf continues to accrete matter and becomes more and more massive. At some time, the dwarf may become more massive than 1.4 solar masses, the limit for stability. At this time, the dwarf becomes unstable and explodes in a type 1a supernova. Because all of these supernovae occur just when the limit of 1.4 solar masses is exceeded, all of them have essentially the same intrinsic brightness. (There are complications, but astronomers have accounted for them by looking in detail at their spectra.) Therefore, these type 1a supernovae can be used as a measure of distance by comparing their apparent brightness with their known intrinsic brightness. We know the intrinsic brightness because we know the distance of relatively nearby supernovae from comparison with other standard candles, such as Cepheids.

We emphasize here that according to the general theory of relativity, the galaxies are not expanding into an empty space that exists independent of matter. It is the universe (space itself) that is expanding. The diameters of individual galaxies are not increasing because matter is sufficiently dense within a galaxy that the attractive force of gravity overcomes the expansion of space. We know this for the Milky Way becuse we observe that the stars in our own galaxy exhibit blueshifts as well as redshifts, without a preponderence of redshifts. If the Milky way is not expanding, then clearly the solar system is not expanding.

On a small scale relative to the size of the visible universe, the universe is very lumpy, by which we mean that matter is not uniformly distributed in space. Rather, matter comes in stars and planets, which are mostly in galaxies. There are some regions in galaxies where the stars are much closer together than in other regions. For example, the nearest star to our sun is about four light-years away, but in some clusters of stars, the average distance between the stars is much less than a light-year. On a larger scale, even galaxies are unevenly distributed in space, with some regions of space containing large clusters of galaxies and other regions, called "voids," being largely empty of any galaxies at all.

However, on still larger scales, measurements have shown that any region of space contains approximately the same number of galaxies. Because of this fact, we say that the universe is "homogeneous" on the average. Furthermore, no matter in which direction we look, on the average the universe looks pretty much the same on very large scales. We say that the universe is "isotropic" on the average, which means that it is the same in all directions. Calculations of the expansion of the universe are much simpler if the assumtions of homogeneity and isotropy are made. The calculations are done within the framework of the general theory of relativity.

Einstein's general theory (without a cosmological constant) admits three kinds of solutions for an expanding universe that is both homogeneous and isotropic. In all three kinds of solution, the gravitational attraction of the matter in the universe slows the expansion as time goes on. But the details of the three kinds of solution are very different.

The first kind of solution occurs if there is a sufficient amount of matter in the universe to halt the expansion and reverse it, so that the universe eventually collapses into a hot dense region, an event called "the big crunch." In this case, according to the theory, the universe is finite in size and has a curvature analogous to the curvature of a spherical surface, except that it is a curvature in three space dimensions rather than on a two-dimensional surface.

The second solution occurs if there is exactly enough matter to keep the universe expanding indefinitely, but if there were any more matter, the universe would eventually collapse. In this case, according to the theory, the universe is infinite in size and is flat, analogous to the surface of a plane but in three spatial dimensions. The amount of matter that makes this situation occur is called the "critical" amount.

The third solution occurs if there is less than the critical amount of matter in the universe. Then the theory says that the universe would expand faster than in the critical case, but would still keep slowing down because of the attraction of gravity. The theory says that again the universe is infinite in size but not flat. Instead, the universe has somewhat the shape of a horse's saddle, except that it doesn't have edges and is in three spatial dimensions.

In all three cases, the theory applies to the entire universe. The visible universe remains finite in extent. The next question is obvious: Does our universe fit any of these three solutions to Einstein's equations? The answer turns out to be that obervations indicate that the visible universe appears to be approximately flat, but the way it is expanding appears to be different from what is expected for a flat universe. We discuss the reason for the anomalous expansion later in this chapter. But before that, we discuss several other things about our universe.

19.2 The cosmic microwave background radiation

Although the simplest explanation for the redshift of distant galaxies is the big bang, not all astronomers accepted that explanation at first. We have already mentioned that the astronomer Fred Hoyle did not believe in it. He agreed that distant galaxies were receding from us, but he and others postulated that new matter was continually being created everywhere in space. The new matter gradually combines into nearby galaxies that replace the galaxies going away from us. Thus the matter in the universe would not get rarer as time went on, but rather the universe could exist forever in a steady (but not static) state.

Recall that microwave electromagnetic radiation has wavelengths shorter than radiowaves but longer than visible light and so is invisible to our eyes. It is radiation similar in wavelength to that emitted by a microwave oven. The discovery of the cosmic microwave background radiation in 1965 by Arno Penzias and Robert Wilson put the steady state theory to rest, for reasons we discuss in the next paragraph. These astronomers were using a large antenna (a radio telescope) to measure the radio-frequency electromagnetic radiation emission from an object in the sky. They discovered that they could not get rid of radiation in the microwave region, no matter where they pointed their antenna. Because this radiation was not what they were looking for, they called it "background radiation." It is called "cosmic" because it apparently came from everywhere in the universe (the cosmos). The average frequency of the radiation corresponded to a temperature of 2.7 K (2.7 degrees above absolute zero), and this temperature was uniform in all directions to the sensitivity of their detector, which had a precison of 0.1 per cent.

Penzias and Wilson did not understand the significance of their discovery until they contacted Jim Peebles of Princeton, whose group was trying to measure just this background radiation. The Princeton group knew that a background radiation would occur as a leftover from early in the big bang, but if the radiation existed, it was incompatible with the steady state theory. Penzias and Wilson and the Princeton group published separate papers on the subject in 1965. As it turned out, twenty years earlier George Gamov, Ralph Alpher, and Robert Herman had predicted that a background radiation from outer space woould be a leftover from the big bang. We discuss the big bang in more detail in the next section.

If a collection of photons is at a certain temperature, it means that the photons have a variety of different energies corresponding to a black-body spectrum characteristic of the particular temperature. Penzias and Wilson could not observe the entire spectrum because part of it was obscured by absorption and emission in our atmosphere. In 1989, a probe called the Cosmic Background Explorer (COBE) was launched into space from a rocket, having the specific task of making precise measurements of the cosmic background radiation. In the next few years, COBE took measurements of nearly the entire spectrum, obtaining a temperature of 2.725 ± 0.002 K. Furthermore, tiny fluctuations in the temperature were found in different regions of the sky.

19.3 The big bang

Our present view of the big bang is as follows: A little less than 14 billion years ago, what is now our visible universe was concentrated in a tiny hot region that blew up. Shortly after the expansion began, the gas consisted of elementary particles, which might have been the quarks, gluons, photons, and leptons we now know (and their antiparticles) plus any other particles (and antiparticles) that might exist but that we have not yet observed. Alternatively, the particles we now know might not be elementary, and the original particles of the big bang might have been what the particles we think of as elementary are composed of. Very early on, the universe was dominated by radiation. This domination means that much more energy was in the photons (and possibly gluons) than in matter. What the universe was like at the instant of the big bang is not known and may remain unknown. What the universe was like before the big bang (if time even existed before the big bang) is also shrouded in mystery.

The entire universe may be much larger than our visible universe.

In some theories the entire universe is infinite in size. Furthermore, in these theories, at the time of the big bang the universe was already infinite, although the part of the universe that is visible today was originally tiny. Of course, we do not know experimentally what is beyond our visible universe, but if a theory about the universe is compatible with what we observe today then we have some confidence that the theory may give us a clue about what is beyond what we can observe. Cosmologists have fertile imaginations and freely speculate about what we cannot observe.

On the scale of the very small, we also make theories about objects that are not observed. For example, we have never observed free quarks, but today most physicists believe in the existence of quarks. There is a difference, however: indirect evidence for quarks bound in hadrons has come from the scattering of hadrons (protons) by electrons, but we do not have even indirect evidence of what lies in parts of the universe that we cannot observe at all.

As the universe (or part of it) expanded, it cooled. As it cooled, the quarks combined into hadrons (mostly the lightest baryons: protons and neutrons). At present our visible universe apparently contains an overwhelming abundance of baryons (matter) compared to antibaryons (antimatter). In our own galaxy, if there were an appreciable number of antibaryons, their annihilation with baryons would produce spectacular amounts of energy that we do not observe. So we conclude that the Milky Way contains baryons rather than antibaryons. The density of matter in other galaxies is such that we conclude that any galaxy contains either matter or antimatter but not both. (There can be a tiny amount of antimatter in a galaxy composed of matter, and vice versa.) In other parts of the universe we observe galaxies colliding with each other. If one of these galaxies were composed mostly of matter and the other of antimatter, we would observe the annihilation energy, which we don't. Although we cannot rule out the possibility that some distant isolated galaxies are mostly antimatter, it seems simplest to suppose that our visible universe contains essentially no antimatter.

The simplest assumption is that shortly after the big bang, the universe (or the part of it that underwent the big bang) contained equal numbers of particles and antiparticles. If the big bang was originally energy that materialized into particles and antiparticles, the numbers of each would be very nearly equal unless the physics at the huge energy density was very different from the physics observed in our universe today. Thus, the simplest assumption is that shortly after the big bang, the universe contained equal numbers of particles and antiparticles.

So if the matter and antimatter in the universe is now matter dominated (as opposed to being equal parts matter and antimatter), it is plausible that some interaction in the early universe did not conserve baryon number and led to a preferential annihilation or decay of antibaryons (or antiquarks) as opposed to baryons (or quarks). Recall that the conservation of baryon number says that the number of baryons minus the number of antibaryons remains constant as time goes on. (Even when the conservation law holds, a baryon and an antibaryon can be created or annihilated at the same time, but the conservation law would forbid an excess of baryons developing.)

After the quarks combined into baryons, further expansion and cooling led to a universe in which most of the ordinary matter was composed of protons, electrons, and neutrons. After sufficient cooling (about three minutes after the big bang), some of the neutrons combined with protons to form helium nuclei, which comprised about 25 per cent by mass of the baryonic matter in the early universe. Other light elements were also formed, but they were less than 1 per cent of the baryonic matter. Neutrons that did not combine with protons decayed into protons, electrons, and antineutrinos. The process of protons and neutrons combining to form nuclei is called "nucleosynthesis."

For more than 300,000 years after the big bang, the temperature was sufficiently high that the electrons and nuclei were ionized into a plasma. (A plasma is an ionized gas.) This is because the universe was sufficiently hot to prevent the positively charged nuclei from capturing the negatively charged electrons. During this time, the photons were continually scattered by charged particles of both signs, and so could not travel freely through the universe. After sufficient cooling occurred, at a time estimated as about 380,000 years after the big bang, the nuclei and electrons were able to combine into neutral atoms, and the photons were able to move freely throughout the universe. This time is called the time of last scattering, and occurred when the temperature of the universe was about 3,000 K.

As the universe continued to expand and cool, the photons were redshifted (cooled) until today they are at the very cold temperature of 2.7 K. This is the temperature of the cosmic microwave background radiation that is observed at the present time.

Today visible matter in the universe is concentrated in galaxies and clusters of galaxies. There are other regions of the universe far from galaxies that have very little visible matter. We often call these regions "voids," although they are not completely empty. At very early times, it is likely that the universe was much smoother than it is now. But the small fluctuations in density of matter that existed in the early universe were magnified by the effects of gravity, which caused more and more clumping as time went on.

It is estimated that the first stars and galaxies condensed from the cosmic dust less than one billion years after the big bang. These first stars contained principally the hydrogen, helium, and other light elements in trace amounts formed by early nucleosynthesis. Any planets that formed around the early stars were unsuitable for the existence of life as we know it because there was not enough carbon, oxygen, and metals such as iron in existence then to support such life.

Stars burn their hydrogen into helium for millions or billions of years. Eventually, however, the hydrogen in a star is used up and the core of the star starts to collapse because the outward pressure of hydrogen burning is no longer available to counteract the inward pull of gravity. As the collapse occurs, the particles move faster and faster, or in other words, the star heats up. With hotter core temperatures, helium burning occurs, and heavier elements are produced, such as carbon and oxygen.

In very large stars, elements can be formed up to iron, which has the largest binding energy of any element. Elements lighter than iron can fuse, releasing energy, and elements heavier than iron can lose some of their protons and neutrons or even undergo fission, again releasing energy. If a star is massive enough, its core contains a large amount of iron. Because iron cannot burn with the release of energy, the star collapses and heats up. This heat is enough to break apart some of the iron, which occurs with the absorption of heat, hastening the collapse. If the collapse occurs very fast, a rebound shock wave occurs, and the star explodes in a supernova.

Some of the early stars were much more massive than the sun, and they ended their normal existence in supernovae explosions. Luckily for us, the conditions in a supernovae explosion are such that many heavy elements are formed. These elements are spewed out into space and mix with clouds of dust between stars. Then, when parts of these clouds condense into new stars, these stars and any planets around them contain more of the elements necessary to sustain life.

Our Milky Way galaxy is estimated to be at least 11 billion years old, and much of the dust in it contains enough heavier elements to support life under the proper conditions. Of course, this is obviously true for our own sun and its planet earth.

19.4 Inflation

The cosmic microwave background radiation leads to a puzzle, which we shall first describe and then discuss a proposal made to solve it. The puzzle is that no matter in which direction in the sky one looks, the microwave background radiation is the same to about one part in 100,000. There are two exceptions to this statement. First, in the plane of the Milky Way, the radiation coming from the galaxy swamps the cosmic microwave background. Second, the earth is moving with respect to the local rest system of the universe. This fact means that the microwave background appears warmer (a slightly higher temperature) in the direction the earth is moving and cooler in the opposite direction. The effect on the temperature of the microwave background is only about 0.1 per cent and corresponds to the earth moving, on average, about 600 km/s (about 400 miles/second) relative to the local rest system in which the cosmic microwave background is isotropic. Because it is believed that the earth is moving relatively slowly within the Milky Way compared to the speed at which the Milky Way is moving, our whole galaxy is moving approximately 600 km/s through the universe.

We have noted that the universe is about 14 billion years old, and the cosmic background radiation originated relatively shortly after the big bang (380,000 years is short compared to 14 billion years). So the cosmic background radiation originated about 14 billion years ago, and in that time traveled a distance of about 14 billion light-years. Now consider the background radiation reaching us from opposite sides of the visible universe. The radiation has been able to travel only half way across the visible universe and so reach us, but it has not been able to travel across the entire visible universe. As we have already noted, no signal can be sent from one place to another in the universe faster than the speed of light. This means that opposite sides of the universe are not now in causal contact with each other. An analysis of how the expansion occurred in the conventional big-bang model tells us that at no time since the particles came into quasi-equilibrium, were opposite sides of the universe in causal contact. The question then arises: How did the cosmic background radiation coming from opposite sides of the universe happen to be at the same temperature if opposite sides could not influence each other?

A scenario invented to solve this problem is known as "inflation." According to the inflation scenario, a fraction of a second after the big bang, the expansion of the universe (or at least part of the universe) stopped slowing down and started speeding up. This speeding up increased the size of the inflating part of the universe by many fold in a tiny fraction of a second, and caused the universe to expand faster than the speed of light. (The theory of general relativity says that no signal can travel faster than the speed of light *within* the universe, but does not forbid the universe itself from expanding faster than light.)

The idea of inflation is a scenario rather than a specific model because, even if inflation is correct, nobody knows what caused inflation to begin or what caused it to end. It has been often assumed that some kind of scalar field with energy was responsible for inflation, but nobody has a good model of this field.

As a result of the rapid inflationary expansion, a region that is now far bigger than the visible universe was originally in causal contact, enabling interactions between particles to cause the entire region to come to quasi-equilibrium. Therefore, the cosmic microwave background could be at the same temperature in all directions. Another consequence of inflation is that the universe was so smoothed out during the inflationary period that the universe should now be flat. This flatness has been approximately confirmed by observations interpreted by theory.

The inflationary scenario has other predictions as well. Although at the time of inflation, the universe was very nearly homogeneous and isotropic, small quantum fluctuations in the energy should have existed. Fluctuations of the magnitude suggested by inflation have been observed by the Cosmic Background Explorer (COBE) and observed in more detail by a subsequent space probe, called the Wilkinson Microwave Anisotropy Probe (WMAP). These small fluctuations in energy were magnified by gravity so that we should have a lumpy universe today that could be the one that we observe.

19.5 What is the universe made of?

We have remarked that, according to general relativity, the universe could be spherical in shape, flat, or shaped like a saddle, although in three dimensions rather than two. Observations, as we have noted, lead us to the conclusion that the universe is flat.

But there is a problem with the flatness: the amount of ordinary matter (called baryonic matter) in the observed universe is far too small to lead to a flat geometry. In fact, the amount of matter visible in the stars and galaxies is estimated to be less than one per cent of the amount required to make the universe flat. Rather, the universe should be open, curved like a saddle, and should expand forever.

But what about baryonic matter that is not in stars but in dust and condensed bodies too small to be luminous? Astrophysicists have estimated the amount of such nonluminous baryonic matter by looking at nucleosynthesis. The ratio of the amount of helium to hydrogen in the universe depends on the amount of baryonic matter there was in the early universe. Calculations using the observed amount of helium as input lead to the estimate that the density of baryonic matter in the universe is about five per cent of the amount necessary to make the universe flat. The conclusion of the astrophysicists is that there is several times as much nonluminous baryonic matter as luminous matter.

But there must be additional dark (nonluminous) matter in the universe in order to account for the motion of stars within our galaxy and the motion of clusters of galaxies. From the speed at which the stars in our galaxy and galaxies in clusters are moving, it was estimated that the additional dark matter contributes about 25 per cent of the amount necessary to make the universe flat. This additional dark matter cannot be baryonic, or it would lead to the wrong amount of helium in the universe.

What, then, is the nature of the additional 25 per cent of dark matter? Although there have been speculations about what it is, nobody really knows. If the extra dark matter does not exist, something must be wrong with our understanding of gravity. Most astrophysicists are reluctant to propose such a radical solution to the problem because there is no strong evidence that the law of gravity fails.

But with 5 per cent baryonic matter and 25 per cent dark non-baryonic matter, we still have only 30 per cent of the matter required to make the universe flat. The apparent solution to this problem was discovered in the 1990s from observations of type 1a supernovae.

Recall that type 1a supernovae are standard candles, enabling us to measure their distance by how faint they are. It has been found that relatively nearby supernovae have redshifts approximately proportional to their distances, as Edwin Hubble first discovered. (This proportionality is called Hubble's law.) But very distant supernovae appear dimmer than they ought to be considering the redshift of their light. This fact was discovered by a systematic study of type 1a supernovae at large distances from the earth. If both the redshift and distance measurements are accurate, the most plausible explanation is that the galaxies with the supernovae were moving slower in the distant past. Let us assume that this explanation is correct.

The argument goes as follows: If the universe expands at a steady rate, then a galaxy going twice as fast as another galaxy will get twice as far away in a given time, in accordance with Hubble' law. But as we look farther away, we look back in time. If the universe expanded more slowly in the past, then a galaxy twice as far away would be going less than twice as fast. In order to find a galaxy going twice as fast, it would have be more than twice as far away. Therefore, a supernova in that galaxy, being farther away, would be dimmer than expected if the universe were expanding at a steady rate.

This conclusion goes against the fact that both ordinary and dark matter are attractive under gravity, and, as time went on, matter should have slowed the down the stars that became supernovae, not speeded them up.

One possible solution to the problem can come from Einstein's cosmological constant. Recall that Einstein added this constant to the equations of general relativity in order to make the universe static, which at the time he believed it was. But it would take a very special value of the cosmological constant to make the universe static. With any other value, it would cause the universe either to slow down more or less than it otherwise would or even to speed up, depending on the sign and magnitude of the constant.

The cosmological constant can be considered as a kind of energy in empty space, often called vacuum energy. If it is positive in sign, it has the effect of acting like repulsive gravity, acting to push the universe apart at ever increasing speeds. Because the universe seems to be speeding up, the vacuum energy must have a greater effect than the matter in the universe, which acts to slow down the expansion. If the vacuum energy is 70 per cent of the energy in the universe, compared to the 30 per cent of the energy in matter (both baryonic and nonbaryonic), then, not only will the expansion speed up, but the total energy in the universe will be 100 per cent of the energy needed to make the universe flat. Thus, although the universe appears to be flat, the way it is expanding appears very different from the way a flat universe without a cosmologcal constant would expand. If the speeding up of the universe continues the way it seems to be going at present, eventually it will expand so fast that distant galaxies would no longer be observable. (At that time, there would be no human beings alive on earth to observe them anyway.)

As the universe expanded in the past, the density of matter became less and less, causing the effects of gravity to become smaller and smaller. On the other hand, if the density of vacuum energy remained constant, then at some time it would become stronger than gravity, and the expansion would speed up instead of slowing down. From the observation of the supernovae data, that time was perhaps about nine billion years ago. Prior to that time, the expansion of the universe was decelerating, except for an apparent brief inflationary time a fraction of a second after the big bang.

If the vacuum energy is constant in space and time, then Einstein's cosmological constant can account for it. However, although the theory of general relativity can accomodate a cosmological constant, the theory does not explain either its magnitude or its sign. But there is another possibility different from a cosmological constant. The vacuum energy might have changed as time went on. A changing vacuum energy is obviously more general than a cosmological constant, and might arise from an unknown scalar field. The vacuum energy is usually called "dark energy," but sometimes it is called "quintessence."

Attempts to calculate the magnitude of dark energy have so far gotten nowhere, as the simplest calculations from quantum field theory lead to estimates far larger than the value observed. So we are left with the unsatisfactory state of affairs in which ordinary baryonic matter contains only 5 per cent of the energy of the universe, while unknown energy, from dark matter and dark energy, contains the remaining 95 per cent. However, we are better off than at the time in the 1920s, when Hubble discovered that the universe was expanding. At that time we did not know that nonbaryonic matter or dark energy even existed. We caution that if any of our assumptions is wrong, for example, if the theory of gravity is wrong, our conclusions about dark matter and dark energy might be profoundly altered.

Chapter 20

Speculations

> There are more things in heaven and earth, Horatio,
> Than are dreamt of in your philosophy.
>
> —William Shakespeare (1564–1616) in *Hamlet*

General relativity and the standard model of elementary particles are very successful theories that describe the way nature behaves. However, as we have said, these two theories do not seem to be compatible with each other. The problem is that general relativity is not a quantum theory while the standard model is. As a result, it does not seem possible to construct a consistent theory of general relativity interacting with the standard model. It appears that there is something we are lacking in our understanding of nature. There have been many speculations about nature that go beyond our present incomplete understanding. These speculations modify general relativity or the standard model or both in such a way as to agree with those theories where they have been confirmed by experiment. In this chapter we discuss onlly two extensions: supersymmetry and string theory.

20.1 Supersymmetry

After Dirac proposed his theory of the electron in 1928, he realized that the theory required the existence of an antiparticle to the electron (a particle with opposite electric charge). At that time, he was very reluctant to propose the existence of a new particle not yet observed in nature. Therefore, he first suggested, as we have already noted, that the antiparticle of the electron might be the proton. However, it was soon realized that, according to the theory, the antiparticle of the electron had to have the same mass as the electron. Because the mass of the proton is almost 2000 times as much as the mass of the electron, Dirac had to withdraw his suggestion. A few years later, in 1932, the antielectron (called the positron) was first observed in an experiment. Dirac's prediction was confirmed.

Pauli was also uneasy with his proposal that an unseen particle, the neutrino, is emitted in some weak interactions. Pauli thought the neutrino interactions were so weak that the neutrino would never be observed. After his death, (anti)neutrinos were in fact observed when an intense antineutrino beam from a nuclear reactor was allowed to strike a large detector. Although nearly all the antineutrinos went through the detector without interacting, occasionally an antineutrino interacted by being absorbed and producing a positron.

Times have changed since the days of Dirac and Pauli. In the late 20th century, physicists were audacious enough to propose that a whole host of as yet unseen particles actually exist, by which we mean that these particles can be created in interactions. All, or nearly all the new particles, would have very short lifetimes, and so would be difficult to observe. The idea leading to the suggestion of many new particles is the notion of supersymmetry. Recall the idea of a symmetry. If a symmetry holds, the results of a certain transformation leave the physical system unchanged. As an example, if we rotate our system to a different angle (with respect to some axis, such as the equator), then the interactions are unchanged. We therefore say that the interactions have rotational symmetry.

Supersymmetry is a symmetry between bosons (which have integral spin) and fermions (which have half-integral spin). If supersymmetry is an exact symmetry, then in the theory an electron, a fermion with spin 1/2, can be transformed into another particle, a boson with spin zero. If the theory describes nature, then the electron must have a supersymmetric partner with spin zero. The theory says that the superpartner, named a "selectron" for supersymmetric electron, should have the same mass as the electron. The result must hold not only for electrons but for all observed particles, including quarks, leptons, and gauge bosons, such as the photon, gluons, and weak bosons. The supersymmetric particles corresponding to the gauge bosons should be fermions. If supersymmetry is applied to general relativity, the resulting theory is called supergravity.

But as of the time of this writing, no supersymmetric particle has been observed. If supersymmetry is a good theory, where are these supersymmetric particles? Why have they not been created in interactions?

The people who have proposed supersymmetry say that the reason supersymmetric particles have not been found is that supersymmetry is a broken symmetry, causing the supersymmetric particles to have much larger masses than ordinary particles. We have already discussed hidden symmetries, also called spontaneously broken symmetries, and most of the proponents of supersymmetry claim that it is spontaneously broken. As a consequence, these people claim, the supersymmetric particles have masses so large that present accelerators do not have enough energy to produce them. This state of affairs will change, they say, when the giant accelerator, the Large Hadron Collider (LHC), comes into operation. As we have said, the LHC is under construction near Geneva, Switzerland. The LHC is designed to accelerate protons to an energy of 7 TeV (trillion electron volts) in opposite directions in two rings and to collide the protons with a total combined energy of 14 TeV. If things go according to schedule, physicists will begin to do experiments with the LHC in 2008. Perhaps some time after that, supersymmetric particles will be observed.

The standard model is not a supersymmetric theory, so if the standard model is correct, supersymmetric partners to the observed particles simply do not exist. When physicists have made calculations with the standard model, so far they have obtained predictions in agreement with experiment, except for certain experiments with neutrinos. The standard model neglects neutrino masses, in contradiction to experimental results with neutrinos, which demonstrate that they have very tiny masses.

So if the standard model gives such good predictions, why do any physicists suggest that supersymmetry has anything to do with nature? There are various reasons. First, in the view of some physicists, the theory of supersymmetry is beautiful, and nature ought to take advantage of this beauty. Dirac was one of the greatest physicists to emphasize beauty as a requirement for a true theory, but he did not suggest supersymmetry, probably because he did not think of it but perhaps because he did not think it beautiful. Beauty in theoretical physics, like other forms of beauty, is in the eye of the beholder.

Another reason why some theoretical physicists are in favor of supersymmetry is quite complicated, but we shall try to explain it. Every interaction we know, including the strong, electromagnetic, and weak interactions, is governed by a number that describes the strength of the interaction. These numbers are often called "coupling constants," but to do so is somewhat misleading. It turns out that the effective strength of an interaction depends on the energy at which the interaction occurs. To take this fact into account, physicists sometimes call the coupling constants "running coupling constants," because they change their values (run) with changing energy. At low energies, the running coupling constants for the strong, electromagnetic, and weak interactions are quite different. However, as the energy is increased, the strengths of the interactions move in different directions. If we extrapolate these strengths to extremely high energies theoretically (the energies are much higher than have been achieved in the labroatory), the coupling strengths of the different interactions tend to become more nearly equal. If at some very high energy, all the interaction strengths become strictly equal, then we have what is called a "grand unified theory," with effectively only one independent coupling strength (or coupling constant). The name grand unified theory is an exaggeration, because the theory does not include gravity.

The point of all this is that the way the coupling constants run with energy theoretically depends on what particles the theory contains. If the theory contains supersymmetric particles with rest energies around a few hundred GeV, then the extrapolation of the three couplings leads to their being the same at the huge energy of about 10^{16} GeV. Thus, the existence of supersymmetry makes plausible the existence of a grand unified theory of the strong, electromagnetic, and weak interactions. Of course, at the present time, the existence of a grand unified theory, like supersymmetry itself, remains a speculation.

But supersymmetry and grand unified theories are not necessarily tied together. There are models of supersymmetry without grand unification, and there are models of grand unification without supersymmetry. It remains to be determined whether both, one, or neither of these theories are realized in nature.

20.2 Superstrings

The basic idea of a theory of strings is that the elementary particles of the standard model are not elementary at all. Rather, these particles are different vibrations of tiny strings, which are too short to have been observed. The strings are called "superstrings," a term that means that the string vibrations include both bosonic and fermionic vibrations so as to include both the bosons and fermions of the standard model.

Two features of superstring theory that have attracted wide attention. The first has to do with gravity. General relativity has resisted attempts by theorists to make it into a quantum field theory. Let us examine the apparent reasons for the failure. General relativity deals with the properties of space and time. In a quantized version, at very small scales, the fabric of spacetime undergoes apparently uncontrollable quantum fluctuations, which become infinite at the positions of the point particles. The methods of getting rid of infinities that arise in other quantum field theories just do not work with quantized general relativity. However, if particles, including the graviton (the quantized gravitational field), are vibrations of strings, their sizes are bigger than zero, and infinities do not arise. It has been a triumph of superstrings that a quantized version of gravity appears to be encompassed by the theory. In addition to a vibrating graviton, the theory includes the graviton's supersymmetric partner, called a gravitino.

The second remarkable feature of superstring theory is that some versions of it seem to make sense only in ten spacetime dimensions. We are of course aware of only four spacetime dimensions—three of space and one of time. Where are the others? The most common explanation given is that the additional six spatial dimensions are not flat but curled up into tiny regions so small that we are not aware of them. Whether we can ever observe the extra dimensions (if they exist) is not known.

There is another popular explanation for why we do not observe the extra dimensions. Subsequent to the introduction of superstring theory, it was realized that there can be other types of structures in addition to strings. A string has only one dimension, length. (Of course, an ordinary string, like a string to wrap a package, has a thickness, but the strings of superstring theory do not.) There also might exist structures with more than one dimension, like the head of a drum, which has two dimensions (neglecting its thickness). Such a structure is called a "membrane" or "brane" for short. There can also in principle be branes in more than two dimensions, especially in a ten-dimensional world. Perhaps the reason we are unaware of the extra dimensions is that we are confined to a three-dimensional spatial brane out of the nine spatial dimensions of the actual world. In this scenario, all the forces of nature except gravity are confined to the three-dimensional brane, whereas the gravitational force can move throughout the larger-dimensional space. In some of these scenarios, the larger space has more than three dimensions but fewer than nine.

Because physicists have thought of more than one string theory, it has been suggested that the different string theories are just different limits of a more fundamental theory, which has been called "M theory." This M theory exists not in ten but in eleven spacetime dimensions. The peculiarity of M theory is that although it might in principle exist, nobody has been able to write down equations that describe it. Therefore, I do not consider M theory as a theory at all but merely an idea or a speculation for which there is no evidence at this time.

If a fundamental theory in eleven dimensions exists, then there are apparently many ways of reducing it to the universe we observe. It also seems there are ways of reducing it to other universes with quite different properties from our own. How, then, are we to predict our own universe among all the ones that seem to be possible.

The answer some physicists give is that nearly all the different universes have properties that cannot support life. Only our universe, or one very similar to ours, can support life, and that is why our universe has to be very much like it is. This idea is called "the anthropic principle," and is much older than string theory. Physicists are divided on whether the anthropic principle is at all useful or whether it is even part of science. Saying that the anthropic principle insures that things must be like they are is not predicting anything.

Nevertheless, although string theory (including theories of branes) has not yet led to any predictions confirmed by experiment, it is too soon to say that it is impossible for future experiments to agree with future predictions. So there is still a chance that string theory is a scientific theory and a chance that nature will agree better with it than nature agrees with general relativity and the standard model.

We have presented in this book a description of the universe as we now understand it, and we have tried to describe what we don't know as well as what we do know. It is sobering that only about five per cent of the universe appears to be understood in terms of the elementary particles of the standard model. The remainder apparently consists of dark matter and dark energy, about which we know very little.

Nevertheless, in the theories of general relativity and the standard model, we have descriptions of matter and energy that appear to hold very well. We understand the forces that enable elementary particles to combine into atoms, molecules, and larger structures. We also have a rough description of the evolution of the universe from shortly after the Big Bang until today, with its complex structure of stars, galaxies, and clusters of galaxies. We know the universe is expanding.

Most of these ideas about elementary particles and the universe as a whole have only emerged in the twentieth century, which was a remarkably fruitful century for understanding physics at very small and very large scales. We await with eager anticipation the future discoveries of the twenty-first century.

Bibliography

[1] Greene, Brian, *The Elegant Universe*, W.W. Norton, 1999.

[2] Greene, Brian, *The Fabric of the Cosmos*, Alfred A. Knopf, New York, 2004.

[3] Guillemin, Victor, *The Story of Quantum Mechanics*, Charles Scribner's Sons, New York, 1968.

[4] Hudson, Alvin, and Nelson, Rex, *University Physics*, Second ed., Saunders, Philadelphia, 1990.

[5] Kane, Gordon, *Supersymmetry: Unveiling the Ultimate Laws of Nature*, Perseus, 2000.

[6] Krauss, Lawrence, *Quintessence*, Basic Books, New York, 2000.

[7] Lederman, Leon with Dick Teresi, *The God Particle*, Houghton Mifflin Co., Boston, 1993.

[8] Lincoln, Don, *Understanding the Universe*, World Scientific, Singapore, 2004.

[9] Newton, Roger, *From Clockwork to Crapshoot: A History of Physics*, Belknap Press of Harvard U. Press, Cambridge, 2007

[10] Smolin, Lee, *The Trouble with Physics*, Houghton Mifflin, New York, 2006.

[11] Snow, Theodore, *The Dynamic Universe*, West Publishing Co., St. Paul, MN, 1983.

[12] Weinberg, Steven, *The First Three Minutes: A Modern View of the Origin of the Universe*, Basic Books, New York, 1977, 1988.

[13] Weinberg, Steven, *The Discovery of Subatomic Particles*, Freeman and Co., New York, 1990.

[14] Woit, Peter, *Not Even Wrong*, Basic Books, New York, 2006.

Index